Modernizing Product Development Processes

Guide for Engineers

Modernizing Product Development Processes

Guide for Engineers

BY

JON M. QUIGLEY
AMOL GULVE

SAE INTERNATIONAL®

Warrendale, Pennsylvania, USA

400 Commonwealth Drive
Warrendale, PA 15096-0001 USA
E-mail: CustomerService@sae.org
Phone: 877-606-7323 (inside USA and Canada)
            724-776-4970 (outside USA)
Fax: 724-776-0790

Library of Congress Catalog Number 2023933976
http://dx.doi.org/10.4271/9781468605426

Information contained in this work has been obtained by SAE International from sources believed to be reliable. However, neither SAE International nor its authors guarantee the accuracy or completeness of any information published herein and neither SAE International nor its authors shall be responsible for any errors, omissions, or damages arising out of use of this information. This work is published with the understanding that SAE International and its authors are supplying information but are not attempting to render engineering or other professional services. If such services are required, the assistance of an appropriate professional should be sought.

ISBN-Print 978-1-4686-0541-9
ISBN-PDF   978-1-4686-0542-6
ISBN-epub   978-1-4686-0543-3

To purchase bulk quantities, please contact: SAE Customer Service

E-mail: CustomerService@sae.org
Phone: 877-606-7323 (inside USA and Canada)
            724-776-4970 (outside USA)
Fax: 724-776-0790

Visit the SAE International Bookstore at books.sae.org

Publisher
Sherry Dickinson Nigam

Product Manager
Amanda Zeidan

Director of Content Management
Kelli Zilko

Production and Manufacturing Associate
Brandon Joy

Contents

Preface

This book is through SAE International, and our experience is largely automotive; as such, this product development work has an automotive setting. That should not suggest that this book only applies to automotive product development. The tools and techniques demonstrated apply to many and perhaps all product development. All product development is learning, and this book demonstrates the questions to pose and mechanisms for learning.

Vehicles have changed over the decades. Products and product development in general have changed. There was a divergence of capabilities and differences in the various vehicle manufacturers. For example, at the outset, there were electric vehicles, steam-powered vehicles, and internal combustion engines. Each manufacturer is betting on a particular propulsion system. Over time there has been a convergence of vehicle platforms along with industry standards and regional regulations.

Some things have stayed the same. Today the vehicle industry finds in the throes of adapting to meet the ecological demands driven by consumers and governmental regulations. Our company will need to understand any emergent technology and the application toward solving a vehicle problem, for example, improving safety or drivability, or meeting a customer's desired feature. It is more complex than grabbing the technology and slapping it on the vehicle. To apply the new technology requires an understanding of that technology and how to specifically apply it. This understanding does not stop at the vehicle but also the application of any emerging technology to the manufacturing lines that produce the components, subsystems, systems, and, ultimately, the vehicle. Development costs and time to market are constant pressures that influence our ability to learn. Prudent decisions are based on experiments, learning and using the appropriate tools to help.

This book follows the product development prerequisites, the things we need to explore up front, and processes and methods to accomplish. The work starts with the need for learning, exploring alternatives, and tools and techniques that facilitate learning. The work then moves to the development and maturity of the product and manufacturing.

Acknowledgments

Jon M Quigley has many folks to thank. At the time of his closing contribution to this effort, his second family's mother, Hilda K. Olive, passed away. Jon was fortunate in his early life to have more than one family. Hilda and George Olive treated him like their own. If he has made anything positive out of his existence, it started with his family and Mr. and Mrs. Olive, Chuck (they played sports together), and Dave (they played music together) and the family of Edward Mendoza.

Jon would also like to thank SAE International for the many opportunities to create new things with them. Specifically, Jon thanks Sherry Nigam, Kayln Karczewski, Amanda Zeidan, Kathy Bratchie, and Monica Nogueira. He wants to thank those with whom he has worked on many other projects, like Roopa Shenoy, Kim Robertson, Martin Rexius, Wesley Chominsky, Shawn P. Quigley, Kenny Thomas, Rick Bertalan, Rick Byrum, and far too many others to mention. Finally, Jon would like to thank all those he has worked with over time at the many organizations he has been with, including those through Value Transformation LLC. Jon prefers to list their names, but that would cover many pages.

Amol Gulve would like to thank first Jon M. Quigley for not giving up on their dream of writing a book which would inspire the young generation of engineers. He would also like to thank the many people who have helped him learn and practice product development processes throughout the years.

Additionally, Amol wants to thank Value Transformation LLC for providing valuable information and resources to write the book. He is also immensely grateful for the continued support of the SAE International team in providing guidelines and critical information to successfully complete the book. Amol could not have done this without your help and passion for sharing the knowledge.

Amol is also immensely grateful to Jon Quigley's leadership and commitment to render a constructive critique on the material, suggestion, or practice that comes with decades of experience working in the automotive and commercial heavy-duty industry. He cannot thank Jon enough for being his unrelenting source of inspiration to challenge how things get done in the automotive world.

Finally, Amol wants to thank his wife, Amruta, for tolerating his incessant disappearances into his home office. A lifelong partner makes both the journey and destination worthwhile.

1

Product Development the Gap

Introduction

Product development (PD) is a crucial collection of functions vital to the success and expansion of a business. It is essential for both national commercial success and global competitiveness. Distilled down, this is no less important for automotive organizations (Figure 1.1).

We can define PD as identifying an untapped market opportunity and converting it into a valuable product for customer delight. We will couple this with emerging technology for the product design and the manufacturing line. Simply put, the PD processes are the mechanisms by which corporations aim to enter or gain in the market by taking the necessary steps to provide products or services (or both) for their targeted customers.

> New product development is the process of putting a new, different product on an existing, new, or growing market. It can be caused by new technologies or new business opportunities.—Eliashberg et al. [1]

FIGURE 1.1   There are many gaps in product development, and effective product development closes this gap.

Customer tastes shift because of fashion, function, or the need to always have something better than what they have now. This something better might be to meet legal constraints, environmental demands, or performance needs. The development of markets across the world is ever changing. Therefore, it is imperative for organizations to continually upgrade and generate new products to keep up with the current demand and meet the wants of their customers. Similarly, it is essential to consider new approaches to manufacturing the product to meet the cost and quality expectations of the customer. Vehicle regulations similarly will impact what product is developed and how we will develop and produce that product.

Creating a new product is an exciting collection of processes that might seem strange to someone who has never done it before. Product creation is part inspiration and exploration, part chaos and discipline. It is creativity and learning, ad hoc and formalism. As a result, a range of skills and talents is required beyond engineering and manufacturing expertise as it exists at any time. Specifically, a collection of emerging markets and technologies present opportunities for the business.

According to Ulrich and Eppinger [2], PD does not just begin and end with introducing a new product. PD is a collection of processes. They emphasize that PD starts with analyzing the market and getting the business perception of the market and then ends with the production of the product, sale of the product, and delivery of the product. Although this is valid, it still fails to describe it completely. It should include the application of new technology. For example, creating a vehicle starts with understanding how to improve the vehicle and manufacture beyond the market changes and segmentation.

As PD begins before we explore specific product incarnations, the PD cycle does not end when a product is released. Since PD refers to the whole lifecycle of a product, companies can continue developing and improving their products for a long time after they initially launch the product.

PD is a broad term for all the steps and goals required to make a new product or improve an existing one. There is no one right way to develop a product because there are many ways to create and enhance a product. It is always situation dependent, although there are some fundamentals. The optimum approach will depend on the organizational structure and competencies, the complexity of the product, and the supply chain required to design and manufacture.

There is a difference between PD and product management. This difference is one of the most significant misunderstandings regarding PD. It is easy to confuse these two names, which relate to quite distinct concepts. Product management is a subset of PD. Product management manages the scope, iterations, and variations, optimizing the product manufacturing.

It is challenging to articulate PD as it is context based and therefore misunderstood since it may apply differently in different industries and businesses. Occasionally, PD mistakenly refers to the development team responsible for creating the proposed solution. In this instance, the PD scope is restricted to a single group, but in our context, it refers to the entire system involved in bringing a product to market.

Product stewardship is a more common term for monitoring the manufacture of a product. This stewardship imparts product managers with the responsibility for managing the solution development, ensuring it has all essential components, and delivering it on schedule in conjunction with the project manager. The product manager is one of the myriad team members responsible for bringing a product to completion in PD. However, it is also a collection of knowledge areas required to articulate and control the development of the product.

In addition to new product development (NPD), with a bit of modification, the above descriptions can describe new service development. Similarly to NPD, new service development includes customer involvement via a customer interface mechanism. The customer interface mechanism works by having customers' opinions in creating a new product or service. In addition, there are specific processes for identifying, managing, and controlling the effort in ways similar to developing the product.

Companies invest a lot of money and resources in creating and marketing new products. Even so, they consider that the new product may encounter unanticipated complications in the development and production processes.

PD activities inside a larger company must consider a stream of different ideas and domains as well as coordinate with existing products and manufacturing, specifically how to choose among these competing priorities and how they have evolved through generations.

Gaps in Product Development

While it is very thrilling, NPD is also quite tricky. Every product and release differs from the ideas, research, and prototypes that go into it. There can be many gaps:

- In product offering.
- In product configuration/variation.

- In market segmentation.
- In PD processes.
- In manufacturing processes (effectivity and efficiency).
- In knowledge to apply new technology to product.
- In expertise to be able to use new technology in manufacturing.
- In post-purchase support trends (aftermarket and serviceability).

The critical factor for successful PD is the harmonization between design, production, market, and product launch. However, from time to time, this harmonization may be disrupted by gaps, specifically in knowledge and capability, that will be required for any specific PD endeavor. Yes, this can include processes.

Gaps in Process

When we talk of gaps in the PD process, we base our explanation on the parts of the market segment that the product is yet to serve. In other words, product gaps involve situations where a product lacks sales opportunities and results in supply outweighing the demand. In addition, there can be gaps in the processes required to manage emerging technology for the product or the manufacturing process. Next, there is continuous pressure on the organization to deliver the product quickly to market (Figure 1.2).

FIGURE 1.2   Market, regulatory, and technology changes impact vehicle development.

lassedesignen/Shutterstock.com.

Gaps in the Knowledge of Emergent Technology

One of the aspects creating a significant gap in PD is the rapidly changing environment. From experience, failures in NPD can be traced back to this ever-changing environment, ranging from market, competition, and technology. Furthermore, this rapidly changing environment ranges from the continuous emergence of new technology and vehicle market segmentation.

A product must be designed to reflect the environmental realities it faces at the time of launch and in the foreseeable future. Hence a company must accommodate a more "adaptive" approach during the PD process. To accomplish this requires a strong understanding of the emerging technology before we start to apply or deploy it.

Companies need to have the ability to respond to new information promptly throughout the developmental process. Dissemination of emerging information and learning is required throughout the project team to adapt and prompt effectively. Dissemination requires tracking individual and organization learning and technological changes in products and processes.

Responding to change and learning new knowledge to innovate often requires relinquishment of obsolete knowledge (Unlearning) to facilitate the creation of new and innovative approaches to the development. —Becker, K

Application of Emergent Technology

An effective organization values continuous learning that includes emergent technology and must apply learning in PD. Organizational learning requires much more than a passing knowledge of emerging technology and the associated context. This knowledge transfer will be valid whether using the technology for the product or manufacturing.

Ideally, the increasing knowledge of emerging technology is applied throughout the PD process. An example of six steps of PD include:

1. Ideation
2. Product planning
3. Designing of product (feasibility)
4. Process of product creation
5. Product testing
6. Commercialization of the product

Product Development Steps

There are many ways to develop a new product; our approach should capitalize on our assets and reduce risks. However, a six-step example for creating and producing a new product is found below.

Ideation: This is the first stage of the process where an organization brainstorms ideas for its following product. In this stage, it is essential to consider factors like the strength, weaknesses, opportunities, and threats (SWOT) analysis, the product functionality, and how it meets customers' expectations (Figure 1.3).

FIGURE 1.3   Example of SWOT template.

Strengths	Weaknesses
Organization internal strengths	Organization internal weaknesses
Opportunities	Threats
Organization external opportunities	Organization internal threats

© SAE International.

Product Planning: After ideation, the second stage is where the businesses pick the best idea for the product and start the plans on how to make it into reality. The factors to consider in this stage would be the value proposition of the product, business analysis, and success metrics.

$$Value = Benefit - Cost$$

Designing of the Product: This is the stage where thorough research and documentation is done on the PD concept. This is where they also come up with a thorough business strategy while also determining the feasibility of the product.

Prototypes in the early phases might range from something uncomplicated like a drawing to something more intricate like a computerized replication of the original concept. Before developing the product, these prototypes enable you to locate potential problem areas.

Process of Product Creation: This is the phase where a tangible product is made.

Product Testing: During this stage, the product is presented to a few people in the market for a pilot test. Here is where the product is carefully analyzed for any mistakes in the functionalities before it is presented to the bigger market as a whole.

Commercialization of the Product: This is the final stage where a product is put into the large market and advertised for customers to purchase and use.

In these six-step processes, the product is developed and manufactured. In the automotive world, we may also use the Advanced Product Quality Planning framework. Each segment of the framework has a specific area of focus.

1. Product design and development
2. Process design and development
3. Product verification
4. Process verification
5. Production

For both examples, the PD start is predicated on our understanding and the stability of the technology. The technology must be stable (predictable) and understood, separate from the development project at least in part. The organization's talent must know how to apply the PD in advance. Not doing so will ensure the project will be lengthy and does not guarantee successful product creation. Separating the learning of the new technology from the application reduces project cost and schedule over-runs.

Business Case

Equally important, will the company benefit from the new product? This benefit might be prestige or income and improved profit margins. Income generating will require an understanding of the business case. Return on investment (ROI) and payback periods are ways to explore the business case.

One mechanism for understanding the business case for the project is the ROI. ROI is a calculation that compares the amount gained against the amount spent as a percentage. The amount earned less paid represents the profit generated from the endeavor. The output is the percent representation of the investment.

$$\text{ROI} = \left(\frac{\text{Amount gained} - \text{Amount spent}}{\text{Amount spent}} \right) \times 100$$

or

$$\text{ROI} = \left(\frac{\text{Profit}}{\text{Amount spent}} \right) \times 100$$

The payback period compares the income stream annually to the expenditure. How many years of production are required to pay the investment? Short payback periods represent a low risk. Longer durations for the payback represent an increase in risk. Returning the investment earlier, the exposure to the loss is more negligible.

$$\text{Payback period} = \frac{\text{Investment}}{\text{Annual cash inflow}}$$

We calculate the future value to compare the expected revenue brought into the company by undertaking the following project (Table 1.1). The value of a dollar today is not the same as a dollar tomorrow. Below is an example that considers an initial investment for a period of six years

$$FV = PV(1+r)^n$$

where
FV is the future value (see revenue chart)
r = interest rate – 8%
n is the compounding period (see years in the chart)

Value = Sum of Present value – Initial cost

Value = 66,930 – 100,000

Net Present Value after 6 years = –33,070

TABLE 1.1 Payback of project investment.

Years	Revenue	Present value
0	$0	-$100,000
1	$9,000	$8,333
2	$12,000	$10,288
3	$15,000	$11,907
4	$20,000	$14,700
5	$18,000	$12,250
6	$15,000	$9,452

In this example, we still would not have recovered the initial investment after six years. So it is not a good investment. But, as always, this is a general statement: estimates can have a high degree of variability and be based on limited information, amounting to educated guesses.

Time to Market of New Technology (Speed)

It takes a certain amount of time to get a product to the market, known as the Time to Market. Time to market covers the entire PD lifecycle, from conceptualization to creation to manufacturing and distribution. The owner of a company where I worked would say, "when can the customer have the product in their hand?" It is the first date the customer can purchase the product creating a revenue stream (okay, maybe it is a revenue creek).

First to market is an advantage. No competition means no substitute product. To that end, companies strive to ensure they are the first to introduce a particular product or feature to the market. Such is the power of time to market; any customer interested in the product has but one recourse to come to the only company offering the product. The company's market positioning highly depends on how soon the product is introduced to the market, among other things.

This idea, often called "Speed to Market," is an essential factor in the success of new or unique PD. Therefore, companies strive to meet the launch timelines and deadlines, which is not easy. Launches of new products or features are connected to model year introductions and high-profile marketing events; for example, there can be a product presentation at the Mid-America Trucking Show for heavy trucks.

Additionally, the company needs to analyze and understand the benefits the target market brings to the table. For example, the estimated potential income generated requires some knowledge of sales price and cost to develop and produce the product. Otherwise, market uncertainty creates doubts about whether the customers will see the benefits of the new product.

Regulatory or legal requirements drive the product developed, for example, emissions or safety products. For these products, customer benefit is secondary, at least. Still others, by the organization's long-term strategy, for example, develop new market segments.

Even if the customer sees the benefits, will there be enough for them to purchase the new product? Unfortunately, studies show that 35% of developed products fail because of pricing market rejection [3].

Pricing is a tricky topic when it comes to PD. A company must determine the most reasonable prices to achieve maximum returns. Doing so is very important for the profitability of the organization. Even so, time to market is an essential factor to consider, bearing in mind that competitor organizations are investing in similar or better inventions.

Delaying the commercialization of your innovations or concepts allows others to surpass you in the market. A delay could make your ideas obsolete since the market advances so quickly. As a result, the optimal price lifecycle for your product or invention keeps shrinking by the day.

Another way to speed up time to market is by finding a market or group of customers where you can test the new or improved products you want to sell with an acceptable level of risk.

They are lead users who will provide valuable input on the quality and use of the product features. Acceptance by the high-end market serves as a barometer for estimating the viability of the innovation in the broader consumer market.

In contrast, the effective dominance of technology cannot be determined until a critical mass of users is attained; specifically, we have a large enough customer base to create a large enough revenue stream to repay the development costs.

Organizations can speed up the time to market by creating well-defined workflows and creating ways of tracking results effectively. The workflow should reduce time wastage and minimize rework and handoff from one department to another.

One way to create a workflow that is well defined in the PD process is by getting rid of the downside of separate groups or departments by setting up a way for everyone to share their knowledge and ideas openly. This method can also be effective in the process of tracking individual results effectively.

The more efficient and productive your company's PD process is, the more accurately you can forecast its time to market. It may also help you plan to launch the product at the appropriate location and perfect time.

Knowledge of the Potential Market

The lack of market research is a current gap in the PD process. Finding new product positioning possibilities frequently involves understanding the set of advantages consumers desire in a specific category and evaluating how current offers by various companies deliver on those benefits.

The competition requires companies to rush to make products because their competitors are making the same product. As much as this idea could be beneficial, it is similarly risky and poses a great danger to the company's returns and reputation.

Below are steps that need to be put into consideration while doing market research before selecting the optimal new product positioning [4, 5]:

1. Identify the characteristics or dimensions that influence a customer's choice in the product category.

2. Get the distribution parameter of customer preference by creating a preference model and estimating it using a sample of customers or sections.

3. Research competitors with the same brand of products and understand the strengths and weaknesses of their best sellers.

4. Use the preference model of the competitor to calculate the market shares for every option under the current circumstances.

5. Check and analyze the performance of any proposed new product.

Development feasibility of new products should be a significant consideration while developing a new product. Technically designing a product is essential in the development process and saves a company time and money.

Technical uncertainty leads to more investment in research and development (R&D) and delayed time to market. Technical assurance of a product is an essential factor to consider in PD because the study shows that 46% of product failures happen at the technical stage (Figure 1.4) [3].

FIGURE 1.4   Successful PD closes the gaps.

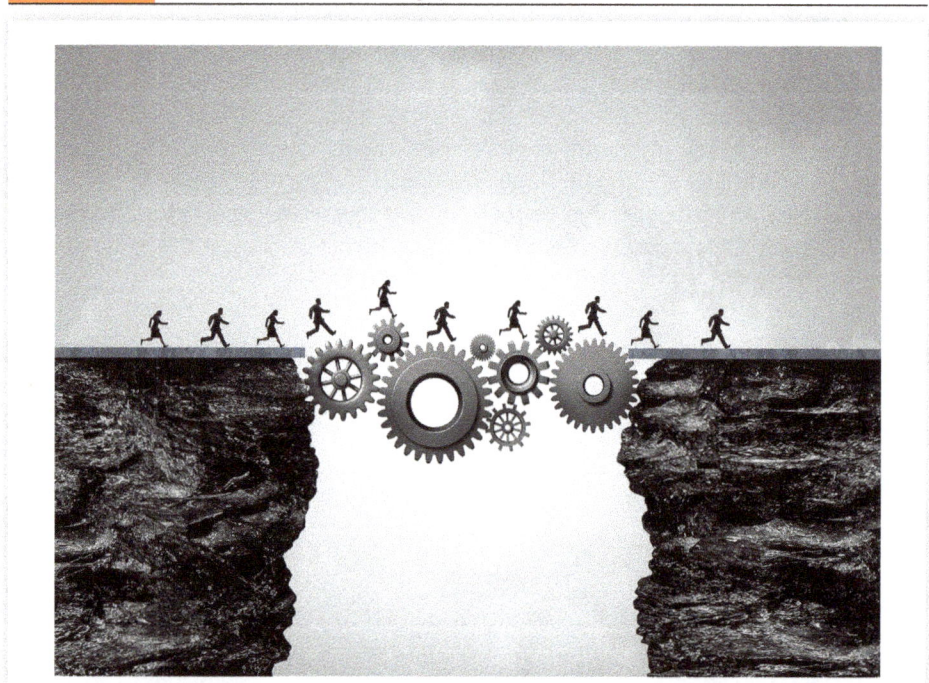

Lightspring/Shutterstock.com.

Ways We Work Now and New Ways to Meet the Gaps

Inadequate communication is a significant current gap in PD. Unfortunately, many factors hinder effective workplace communication. One of them being office politics. Office politics create a division of teams because of the scramble for power. As a result, the division of groups working on PD leads to inadequate communication and senseless conflicts, which then disrupts the harmony of the PD process. In addition, having a perceived political environment in the workplace disrupts the workflow and leads to low productivity.

We can close the communication gap in PD by encouraging workplace teamwork. Ensuring every employee understands that they are working toward the same goals is vital for the smooth running of the PD process. All these are effected by the organization's culture.

Another way we work now would describe how many businesses partition (organizational structure) their innovation and creative centers. As a result, the primary business functions may not be fully engaged in innovation and development. As a result, lower tolerance for risk can slow down the time to market.

A way to meet this gap would be for more businesses to focus on product innovation in the company. This way, more organizational teams can be involved in creating new products (Figure 1.5).

FIGURE 1.5   PD must balance many desired objectives.

EtiAmmos/Shutterstock.com.

Organization and Process Assets

Organizational Process Assets are documents describing specific work approaches acquired by an organization, often over the years—process assets are articulated for projects, products, and quality management. The process assets may include guidelines, policies, assessment tools, or sometimes knowledge gained from experience and lessons learned.

An organization's process assets are essential in PD, especially in the planning stage, regardless of the project. Often these assets result in standards and, ideally, from within the organization. It is sound advice to project managers to use the organization's process assets. A team (and executive) discussion should ensure if alteration or deviation from the organization's process assets is required. Any process change for the project should be clearly articulated. For example, one of us was a process manager for an embedded electrical/electronic department. The first section of the process description defined the specific objective of a process. Why are we doing this? Sometimes the process cannot be executed as is, and sometimes the process does not apply to a particular project because of scope. Sometimes the inputs to the process are not ideal, and we must adapt. Ideally, updating our process assets captures team learning obtained through working with the process.

Supporting Learning the Emergent Technology

Organizations need to support and accommodate learning as part of their business lives. Continuous learning puts top businesses at the highest ranks in the business world. There is more to this than capturing knowledge via processes and work, though this is significant. Organizations can support continuous learning by providing employee training from time to time to keep them updated on the changing world.

They can also invest in sending their employees to business workshops and seminars concerning the field of business in which the organization specializes. Workshops are like business socialization events. Our team members can meet others with experience and knowledge of specific technology and applications.

Workshops are where different businesses exchange ideas and share their success strategies. Aside from that, it is an opportunity for similar companies to network and expand their horizons.

In addition to workshops, there are industry advocates such as SAE International and the standards committees, which often consist of representatives from the original equipment manufacturers (OEMs) and the supply base to define vehicle and technology standards.

Support Application of the Emergent Tech

The purpose of knowledge is action, not knowledge.—Aristotle

It should be evident that after learning something, the next step is putting the knowledge to good use. One of the ways that organizations can support the application of the knowledge acquired is by encouraging the transfer of knowledge from those that have it, facilitating action in the organization at large. One way to do this is via training sessions among employees. For instance, employees can attend internal workshops or seminars ideally conducted by those within the company with the knowledge. This way, those who know can spread the knowledge learned throughout the organization.

Another internal support mechanism is through communities of practice. Communities of practice are the gathering place for those skilled in technology or specific processes. This accretion of talent is then available for individuals and teams as consultancy to any effort or project. Communities of practice can originate organically from within the organization or via executive mandate. In either instance, communities of practice should not end up as a euphemism for outsourcing to another part of the company. To keep the community of practice fresh, we can rotate those on this team (Figure 1.6).

FIGURE 1.6 External changes compel an organization to respond rapidly and effectively.

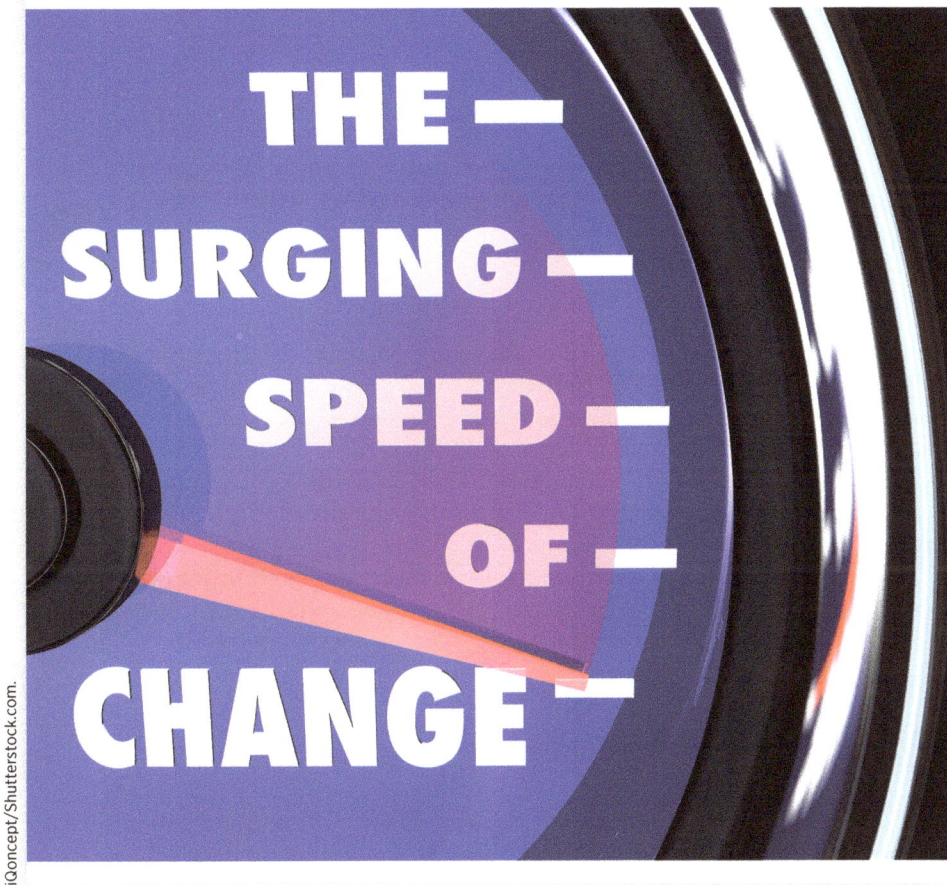

iQoncept/Shutterstock.com.

A Rapidly Changing Environment

Because of the high volatility in the 21st century product markets, PD organizations are faced with the challenge of getting the hang of a fast-moving target. In addition, the frequently changing needs of customers are acknowledged to be a key source of market turbulence, adding to the uncertainty of PD success.

Therefore, there is a need to introduce new products in the market from time to time and the pressure to keep up. Aside from the frequently changing customer needs, a few more factors have contributed to the increased importance of introducing new products to the market.

Some factors include the shortening of lifecycles of products, the increasing globalization, and the increased number of customized products. Therefore, a company's ability to adjust is essential for its survival and competitiveness.

Adjusting to the Rapidly Changing Environment

One way to remain in business and on top of the competition is by beating the competition in NPD. As a result, there is an urgency to keep revising the current products, looking for ways to improve them, and introducing the newly enhanced products to the market as quickly as possible.

> The ability to develop, produce and introduce new products faster than the competition is an important factor for success in companies. —Carrillo and Franza [6]

Planning

The company's success highly depends on the company's ability to be the first to introduce a new product in the market. Companies must speed up the launch of new products and shorten their development cycles because of shorter product lifecycles and increased market competition [7]. First, the market opportunity must be successfully identified while accounting for the short time-to-volume factor. Additionally, these companies must achieve the required production volume while achieving the product and manufacturing performance targets.

Planning is vital for PD. Project planning, product, and manufacturing planning help us prioritize resource use, talent application, and learning. Failure to plan and opting for an entirely adaptive or reactive approach is likely not to yield desired results.

> By failing to prepare, you are preparing to fail. —Benjamin Franklin

Plans are how we believe things can happen. Doing the work informs if there is merit in those beliefs. The organization and any project must be able to adapt to what is learned while doing the work.

> Plans are of little importance, but planning is essential.—Winston Churchill

There is room for justifiable criticism of formal long-term project planning. The critique is that planning takes too much time and goes on too long without metrics and data to support that plan. However, we should not throw the proverbial baby out with the bathwater. It is as easy as constraining the planning to the immediate. As far as we know, there is no dictate on the length and scope of the planning effort. Instead, integrate adapting into the planning approach to resolve these errors in the process.

It takes as much energy to wish as it does to plan.—Eleanor Roosevelt

Recognizing that in some elements of the work there is sufficient understanding to be able to plan, and other factors, much less so, may require a different approach to planning. For example, we can adopt a variable approach to planning rather than a dogmatic approach. Like our earlier process story, the planning approach matches the circumstances.

Data

There are benefits to the organization's process beyond the repeatability. Providing our teams are availing themselves of the methods and accurately recording metrics, there will be data related to those processes, or there should be. One of the best ways to understand what is likely to happen is to review past results.

Progress, far from consisting in change, depends on retentiveness. When change is absolute there remains no being to improve and no direction is set for possible improvement: and when experience is not retained as among savages, infancy is perpetual. Those who cannot remember the past are condemned to repeat it. In the first stage of life the mind is frivolous and easily distracted; it misses progress by failing in consecutiveness and persistence. This is the condition of children and barbarians, in which instinct has learned nothing from experience. —George Santayana

Data derived from the organization's effort, specifically the processes, make accurate estimates and predictions more probable. For example, we can view the distribution of outcomes of a specific step and the range of possibilities if we record our process results. For example, we recorded and tracked the software release process when we were process managers. We discovered that this effort consistently took 32 hours in time and effort.

Talent

To paraphrase a line from scene II of Shakespeare's *Henry V*, the organization's talent is the sinews of the organization's power. Without an engaging collection of talent willing to constantly learn, the organization will not be productive. Team members thus engaged do not happen by accident. The organization's priorities and culture must support such a desire. Organizations invest time and resources into getting feedback from the team and listening to their customer's (internal and external) complaints (Figure 1.7).

FIGURE 1.7 Generic process flow applies internally and externally.

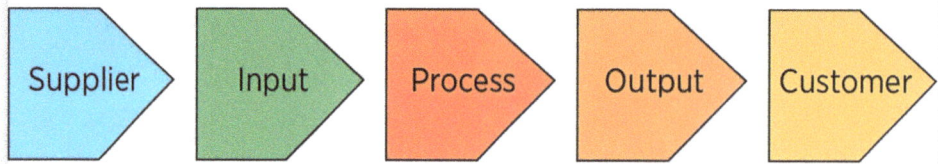

Ideally, the workplace allows for time to learn while doing the work. The company provides coaching and training to keep the team ever improving. As I used to say to my coworker Wes, sometimes the axe must have the blades sharpened. This axe story is an analog for saying our talent needs to keep up with the rapidly changing environment. The workshops are beneficial because, aside from networking, it is also possible to pick up some trendy skills that would benefit the employees' productivity and overall performance.

A productive and innovative team can support and grow a business even in uncertain times. As much as qualifications are essential in a job, the willingness to learn and improvise in the team is what puts one institution in the top ranks of any business field. In our book *Continuous and Embedded Learning for Organizations*, we review the two dimensions of engaging the team, empowerment and agency. An individual is empowered through the organization's expectations, culture, and environment. Agency is internal to the individual; the individual believes they can change their circumstances. Hiring adaptable people for PD is another way to help a business adjust to the rapidly changing environment. Having the best team does not necessarily mean the most qualified team, it means having a competent and innovative team.

Areas of Action

Organizational innovation may pertain to business practices, workplace organization, working methods, or external relations. For example, innovation may involve the creation of databases, the training of personnel, and the implementation of production or supply chain management systems, quality management systems, production systems that integrate sales and production, or the development of new methods of supplier integration.

Marketing innovation involves substantial changes in design or packaging—form and appearance, design and placement—new sales channels or new methods of presentation and exhibition.

Product innovation is introducing a novel product or incorporating a novelty into an existing product. Design and creation can be a source of non-technological innovation that enhances and complements technological innovation. Product innovation, from experience, is closely connected to strategic relationships with the supply base and technical expertise.

Limitations of the Traditional Way of Working

We cannot solve our problems with the same thinking we used when we created them. —Albert Einstein

An organization must continuously seek to improve, understand, and adapt to the external environment while improving for it to remain relevant. No matter how successful a business is, competition does not get easier. Hence, creating an organizational culture of learning, prioritizing learning of the individual at teams, and ways to record and disseminate.

Therefore, to remain in business, even the top most successful companies must constantly find ways to bring value to their customers and stand out. According to Forbes, there is an expected drop in returns on stock market investment from 7% to 3%. Today's success is not tomorrow's success.

What Are the Limitations?

Unmet Customer Needs

Trying to meet customer needs through PD is like steering a large ship. The customer does not immediately witness the result of what we do to meet the customer needs, at least if we do not take action to ensure that our efforts to meet their needs are demonstrated frequently or periodically along the way.

All too frequently, customers are involved after the new product is launched. We ask for customer feedback after investing significant time and effort on their behalf. We might also ask for specific requirements via market studies or surveys at the beginning [8].

We may have a misguided belief that we should aim to satisfy as many individual customers' needs as possible. However, prioritizing the effort this way is not a ubiquitously correct approach. Being dogmatic is seldom a great approach. Some things are more important than others, and focusing on these is the most important. Risking the essential product features or attributes for lesser-value parts is not prudent.

The language of the customer and the development organization may not be congruent. Indeed, the lexicon of engineers and business personnel, and customers are not the same. But this lexicon challenge with both development team members and customers can be distributed across the globe, making defining what is significant a nontrivial challenge.

A different approach is required to bridge this gap in the lexicon, language, and culture. We do this with scenario-based product descriptions, sometimes called use cases. Use cases are actor-based descriptions of the features, often graphically. These graphics clearly articulate the interactions between the customer and the product or system. These simple graphics are the foundation for applying the engineering language, technical definition, and subsequent challenges.

Time Wastage

Traditional PD processes involve hierarchies and approvals at every stage of the development process. Organizations are getting better at this. However, there are benefits to the organization's structure. From experience, an accretion of competency that often comes with functional structures in the organization comes with a growth in competency and tools to facilitate the work.

Competing priorities of the various managers involved in any PD efforts can also lead to wasting time. Political wrestling and individual evaluation criterion from the organization leads to competing priorities of the organization and the individuals.

Escalation plans can help to resolve some of these bottlenecks. For example, one of us once had an internal supplier at the organization where we worked to attempt to push through a gate. The supplier wanted to pass the gate though the requirements were not met. After going to the Vice President (VP) of Quality for the organization and discussing the state of the project-reviewing metrics, the VP said he would audit the next gate for the project. The internal supplier decided not to press the gate date.

Changes in Technology

The application of technology is now experiencing a foundational change. The rate of change, it seems, is ever increasing. For example, vehicles took decades from a few electronic control units (ECUs) for the engine to ever-growing numbers. Now a range of propulsion systems are being explored (battery, hybrid, hydrogen fuel, hydrogen fuel cell), as well as many driver assistance systems.

Organizations have a tradition of being set for handling machine-based technologies. These organizations' primary focus was on the steady and efficient utilization of physical resources like mass production. However, modern industry is progressing to significant automation levels beyond manufacturing. This approach is not productive for knowledge work, which is not deterministic as a manufacturing line but probabilistic. That which we learn is not known until we learn it. If we knew, we would not be required to learn. The organization must be able to meet the challenges presented effectively, and this requires an amalgam of probabilistic and deterministic approaches.

Technology changes impact beyond organizational structures but also the talent of the organization. Our team will need to be continuous learners in the modern organization. It is not just the technology as it applies to the development of the product and end application but also the techniques and technology used to develop the product.

Costly

PD can be costly. Unfortunately, specialized tools (investments) and talent are not without cost and are often over extended periods. Additionally, there is no guarantee that there will be a payback. Learning may require experimentation and exploration without a clear vision of the application of the technology.

No Room for Flexibility

Organizational structures and processes define how to work; procedures are in place to ensure a repeatable work outcome. However, PD is not much like an assembly line—the ability of the organization to adapt to meet external and internal challenges and technology. There are times to be inflexible, and there are times to be adaptive. From experience, PD requires discipline and adaptive approaches.

Poor Quality Products

If you have been in any industry or following it for a while, you will have witnessed quality problems that have badly impacted the customer. The quality and riskiness of the product apply to more than just the automotive industry. This quality pressure compels the organization to document and use formal processes for the development effort.

Generation Dynamics and High Customer Expectations

What is the worst possible result, and at the same time realistic, that a product can have? PD was completed, but the customer ended up unhappy with the outcome, thus putting a dent in the company's reputation.

Typically, this is an occurrence that is often seen when the expectations of the customer are not clearly articulated, or a disagreement arises around the same. When we rush to confront a product or when a corporation wants the work so much that it does not want to raise any worries in the client's mind with too many questions, this is a problem that often occurs from experience.

Many of the challenges that drive us to ignore customer expectations can be fixed with a well-defined and easily replicable onboarding process. Even so, there are a few tried-and-true strategies for keeping everyone on the same page when it comes to managing and setting expectations.

When we talk about customer expectations in PD, we refer to the set of standards that customers anticipate to find in a certain product. Customer expectations may be strategic involving a detailed feedback process while others may be reactive and emotional.

Everything the customer knows about you beforehand will allow them to build their ideas about your way of working, and as the project develops, the relationship will be strengthened or weakened.

And to fulfil it, it is important to differentiate the expectations of your client's needs.

"The customer is always right" is an adage that, in this context, means you must understand and manage their expectations appropriately, even in cases where you thought some of their comments or behaviors were inappropriate, as you probably already have ever lived.

Difference between Customer Expectations and Needs

The customer needs are always brought up while discussing their demands. In other words, the individual has already selected the element they wish to alter and, in many cases, is aware that the change must occur quickly. Accordingly, one could say that a need is a specific target.

Unlike a need, expectations are subjective and can change from moment to moment; in other words, they are influenced by the client's mood and can be unpredictable. Expectations are customer desires.

Why Should You Know and Meet the Expectations of Your Customers?

Some of the benefits that managing your customers' expectations will give you are as follows:

1. You Will Make Your Work Dynamics More Efficient
   Each customer represents a challenge in itself, although usually the most demanding clients are the ones who will make you grow individually and as a team. If you feel like you are in a comfort zone, an excellent way to get to the next level is to recognize what you are not doing well or what requires superior performance.

2. You Will Create a Long-Term Bond
   If you meet their expectations, operationally and in terms of their specific needs (Which media do they prefer to use? How do they like to be treated?), you lay the foundation for their loyalty. This is one of the objectives that should be an essential

part of all your strategies, from product design and marketing to service and operational management.

Think that 56% of customers are more loyal to companies that understand their needs, preferences, and priorities.

3. You Will Educate Your Client

You already know that "The customer is always right," and it is true although that does not mean that you cannot present your reasons and expand their panorama regarding the area in which you are an expert. Maybe they expect your product to give them results overnight, which you know is a pipe dream. Will you be able to respond with arguments and specific information that will help them understand your work, what it is, and how you do it?

4. You Will Offer an Extraordinary Experience

Every person needs recognition to feel valuable, and the same happens with your clients. If they see you strategically responding to their needs, you will not only give them results they like about your product, but you will also encourage a positive emotional response. Thus, interacting with your company will be a delight.

5. You Will Gain an Ambassador for Your Brand

See yourself as a customer of the brands you like the most. What makes you say: "I recommend the designers of agency X, it's great"? Sure, it is because they perform well, but even more so because they knew how to respond to what you needed, even when things were not always perfect.

This is how you will get your clients to recommend you and give your company valuable input from prospects, in addition to strengthening your image in front of the public.

Factors Influencing Customer Expectations

Delve into what are the five factors that influence the expectations that your customers have about the product. Thanks to these factors, you will be able to understand your customers in-depth and what your field of action is to improve their experience.

1. Your Personal Principles and Values

Like you, the client is governed by specific guidelines that determine the value they give to a particular aspect, not only of your product but of any experience. Some people want to be answered as soon as possible; otherwise, they will feel they are not considered. Others will not have a problem regarding response times but prefer that you give them a detailed explanation, even if it arrives hours later. If you don't, they will believe that you do not think seriously about each of your proposals.

Try to understand their ways of thinking and evaluate the actions to follow so that you react similarly to your clients.

2. The Organizational Culture to Which It Belongs

No organizational culture is so simple that it can be immediately categorized, but some are more results-oriented and others more people-oriented. If your customer wants to deliver certain products (remember they answer to someone else in their company), no matter how you get them, you will have to answer the same way.

On the other hand, if they are more interested in creating a productive link because their company's culture has that focus, they will be more willing to learn about your work dynamics.

3. Your Mood

   Everyone has good, bad, and worse days. This means that your client's state of mind dramatically influences their judgments of you and other affective aspects that, although they do not depend on you, are at stake.

   In that case, one of your principles should be to respond actively and empathetically instead of reacting negatively when you think your client is not expressing themselves in a friendly tone.

4. Your Past Experiences with Other Providers

   If you have already passed through three restaurants that offered food without seasoning, you will have a bias in your perception when you go to the fourth. Before they serve you, you will already be thinking that a new disappointment is coming and that you will not hesitate to point out what does not satisfy you.

   Thus, if your clients have suppliers who do not understand or meet their expectations, deep down, they will believe that you will not make any difference. Prove them otherwise and create a unique experience.

5. What They Know about Your Organization

   The quality of your clients' references about your company will largely determine their perception. Having negative reviews online is a risk, but you should also respond to the positive ones. Perhaps when a client put on your Facebook site "This is the best accounting management service I have ever worked with," they had already had a long-term business relationship where they faced and overcame difficulties together. Instead, your new customer might say, "Great! I expect perfect service from day one"; in that case, you should use the information they have about you and manage it as well as possible so that their interactions are valuable and fruitful.

How Companies Can Meet Customer Expectations

There are many ways to manage/meet customer expectations in any product. However, considering a rapidly changing environment, one of the most effective ways to meet customer expectations is through process value analysis. Companies should consider paying more attention to process value analysis in the product generation dynamics.

In the present day, product managers should aim at generating products that satisfy their customers' emotional needs. It is the only way to stand out in a market with increased diversification and too many varieties of similar products. Process value analysis comes in when companies want to provide these products in the market as fast as possible, while also working under a reasonable budget during the PD process.

Process value analysis can be defined as the examination and review of each step of the product generation process to see if each activity cost-effectively ensures customer satisfaction. When the analysis is done, and a certain activity does not provide value to customers, it can be streamlined or improved if possible.

Up until now, manufacturers have been concentrating primarily on product value, which is dependent on how well the finished product operates. However, consumers of the products are proactive and creative and should be given a chance to design the kind of products that would meet their needs.

Behavior economics stresses how important experience value is. Involving consumers in the product generation process would provide them with the experience of actually creating a product of their choice. Such encounters would undoubtedly satiate the deepest human need for a challenge as well as self-actualization.

With such process qualities, value is added rapidly and wonderfully, and all clients are completely satisfied. Process value has the drawback of being difficult to assess in terms of quality. However, it may be resolved extremely well if a pattern classification technique were introduced.

Additionally, such an approach makes product generation highly flexible and adaptable to satisfy the widely different and constantly changing requirements or expectations from our customers.

Product Generation Dynamics that Meet High Customer Expectations

Product generation dynamics involves changes that can be put in the process of PD, to meet customer needs and increase the chances of success during product launch.

The objective of tackling the issue of changing client requirements is to reduce the unpredictability that the factor brings to PD initiatives. So far, a variety of solutions have been presented, which may be divided into three categories:

1. Increasing sensitivity to customer needs changes
2. Reducing the product generation cycle time
3. Forecasting

Increasing Sensitivity to Customer Needs Changes

Aside from listening to customers during market research of PD, product developers should normalize listening regularly enough. Flores [9], in her framework "Innovation by Listening Carefully to Customers," talks about actively listening to the needs of customers more frequently.

The rising unpredictability of customer requirements in the market has necessitated the development of new methodologies and paradigms for keeping tight customer touch. Therefore, ongoing customer satisfaction data collection is thought to be crucial in managing long-term competitive advantage.

Maguire et al. [10] did a study on four successful companies that were achieving successful PD. These companies were deriving their success from having the desire to understand their customer needs at all costs.

They were using customer-listening tools that helped us get up to date with the customer's changing needs throughout the product generation process. To understand the customer's changing needs, a company has to, first, acknowledge that it is a natural process.

Maguire encouraged frequent assessment of customer satisfaction and measurements of the same. Customer surveys can be used as a tool to monitor the fluctuation of the customer needs from one state to another.

Additionally, a tool based on the Markov chain model and quality function deployment (QFD) was presented to monitor fluctuating client requirements using the probability perspective. However, needs that vary quickly may not be compatible with the dynamic QFD, and the implementation is difficult because of the large quantity of input data needed [11].

The fact that client surveys must be manually evaluated and processed can be seen as the lack of autonomy of the system and considered as a restriction. In particular, the technology might inform users proactively when incorrect client needs are being focused on.

According to Griffin A. and Hauser R.J. [12], capabilities for core manufacturing in fast-paced markets should be market driven in that the major focus should constantly be the customers' voices. There have been suggested frameworks that try to make the allocation of production resources more responsive to the shifting demand.

For instance, the concept of agile manufacturing was developed in response to the challenging business climates brought on by the individualized nature of client demands. The framework encourages evaluating changing consumer needs so that flexible manufacturing facilities may adapt to the changes in an efficient manner.

Reichwald et al. [13] came up with a new organizational strategy in their attempt to solve the uncertainty brought about by the rapidly changing customer needs. This strategy was meant to increase sensitivity to the dynamic customer needs.

It was suggested that distributed mini-factory structures take the role of the current job divide between centralized production and dispersed sales. The goal of the mini-factories is to combine manufacturing, sales, innovation, and consumer engagement in close proximity to the customers.

The setup, in their opinion, would allow for real-time management of industrial facilities (in the mini-factories). The suggested paradigm intends to include customers as a crucial component of the dynamic value-added network, allowing responsiveness to rapidly changing demand.

Through supply networks, the ambiguity of end consumers' future demands might spread. Therefore, supply chains should interact physically or online to exchange information of the dynamic consumer demands.

Reed and Walsh [14] pushed for the sharing of technology future plans inside supply chains to increase providers' technological capabilities. Buyers' technological roadmaps should be in line with suppliers' roadmaps as they are more likely to understand end-users' future demands.

Aside from physically becoming closer to clients, the Internet has been utilized to manage customer relationships. Gunasekaran et al. [15] suggested a web-enabled architecture for the continuous exchange of information of consumer demands across a geographically distributed supply chain.

As a result of acknowledging the shifting customer expectations, product quality may be enhanced (Figure 1.8).

FIGURE 1.8 An example of a product lifecycle.

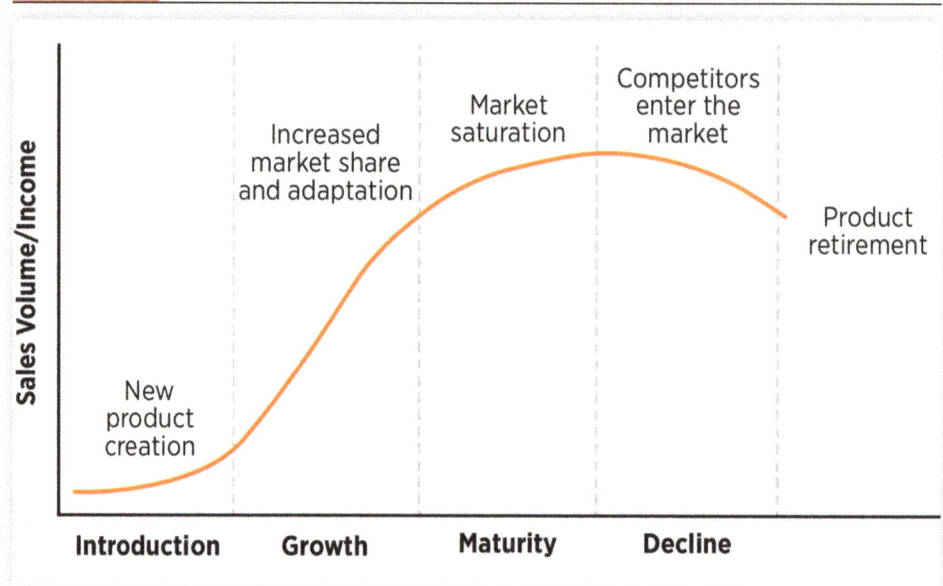

Sales Volume/Income

New product creation

Increased market share and adaptation

Market saturation

Competitors enter the market

Product retirement

Introduction **Growth** **Maturity** **Decline**

Reducing the Product Generation Cycle Time

The elapsed time between when a customer expresses their need and the time of product launch is what is described by the product generation cycle time and should be cut short. The act of cutting short the product generation cycle to gain a competitive edge in the industry is also known as time-based competition.

Reinertsen and Smith [16] emphasized that a market clock begins to run as soon as the first company in the sector starts the development process. As a result, user needs, which are dependent on time, may start to change as the seconds pass.

Reducing the product generation cycle time speeds up the time to market and lessens the impact of rapidly shifting consumer demands (Figure 1.9).

FIGURE 1.9 Faster changing market demand often shortens the product lifecycle.

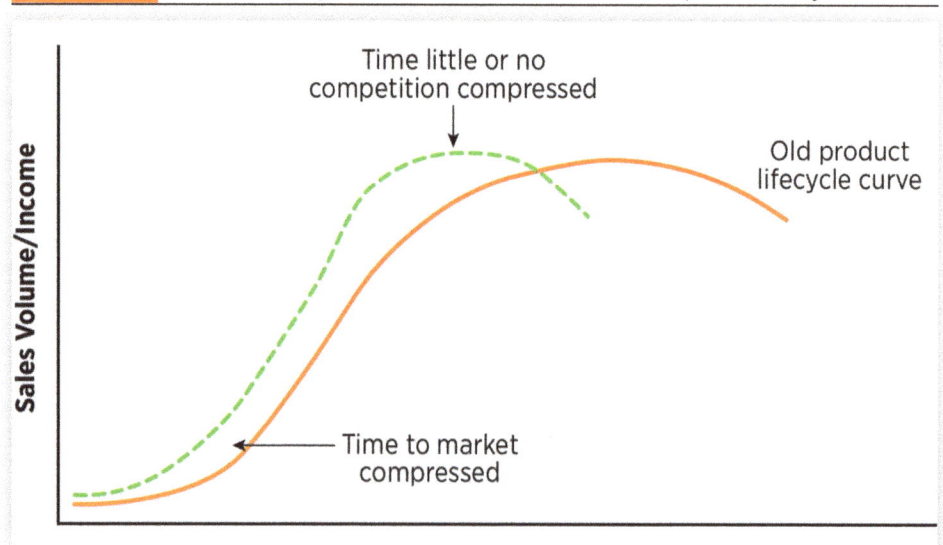

Sales Volume/Income

Time little or no competition compressed

Old product lifecycle curve

Time to market compressed

... in high-growth markets involving short product lifecycle, the overall impact of speed-to-market on profitability is compelling. —Karagozoglu and Brown [17]

Reinertsen and Smith [16] had a recommendation to help accelerate the time to market as a competitive strategy. They suggested that a company can set aside one division to focus on the complete grasp of the market.

Following that most delays in the product generation process and product launch failures arise from inadequate market research and understanding, Chen et al. [18] also came to the conclusion through empirical research that it is more crucial to implement a time-based approach in an untapped, developing, or rapidly changing market than in one that is well established, stable, and well known.

The reduction of the time cycle in the new product generation process gives a high chance of reducing product success uncertainty. —Jetter [19]

Forecasting

The act of forecasting is described as the prediction of events or future market trends in the PD sector. One strategy used to control risk brought on by future uncertainty is the forecasting of variables.

In PD, some of the most forecasted variables include customer needs, purchase intention, customer behavior, and market demand. Market demand has received the most attention among the several factors in the literature on product forecasting.

It is not surprising to see that forecasting on dynamic customer needs is not as expected. Although the customer needs and market demand are related to products, they are distinct pieces of knowledge with various consequences and uses.

The subject variable for estimating market demand is volume, often of items or product categories with certain qualities. The predicting of consumer demands, on the other hand, pertains to product qualities that customers seek, perhaps independently of specific items.

Regarding applications, the former may help with administrative choices like allocating manufacturing resources and developing a marketing strategy, while the latter can provide clear indications to product designers and other functions involved in product creation.

Szakonyi [20] puts forward that marketing organizations should take responsibility for their market completion and do all the work, including forecasting on future customer needs.

A forecasting technique based on a fusion of QFD and S-curve analysis was also suggested by Mclaughlin and Stratman [21]. The QFD matrix is used to compare future customer requirements to the technology prediction performed using the S-curve approach.

Flores [9] was keen on realizing that customers have been observed to try and create a future for themselves in terms of products. Therefore, a careful observation must be made and organizations need to pay attention to where this new strategy is headed to.

Without any doubt, communication was highly encouraged for the purpose of meeting future customer needs and successful product generation. Flores went ahead to recommend a method that encourages businesses to actively hear future consumer concerns.

According to reports, it was embraced by companies like IBM, Sony, and Nike to make it easier to create goods that may satisfy future demands. This kind of predicting is mostly reliant on how clients view the future.

Xie et al. [11] put in application a time-series model aimed at mathematically predicting future customer needs. To be specific, they focused on the double exponential smoothing technique.

The double exponential smoothing technique is an approach accommodating trends that evolve in two ways: addition and multiplication. For it to go either way, it depends on whether the trend is linear or exponential.

However, it should be highlighted that the approach is only applicable to forecasting time-series data with a linear trend. To some extent, the constraint may preclude the approach from being applicable to actual issues where nonlinearity is prominent.

Wu et al. [22] presented the gray theory for use in forecasting customer requirements. The gray theory required only four historical data points.

In contrast to other time-series approaches which require additional data points, this one does not. The necessity for human interaction to update both freshly emerging and out-of-date consumer needs throughout the timeline is a prevalent shortcoming of these systems.

These systems applications in rapidly changing markets would benefit from solutions that can autonomously evolve the pertinent set of consumer needs.

Chen and Yan [23] used the radial basis function neural network to examine and analyze the utility of customers. They did this using a moving window architecture which proved that the method may reflect and foresee changes in customer desire for a collection of design proposals.

With the comparison of projecting future consumer demands, the technique generates anticipated client preferences over a variety of design possibilities.

Conclusion

To bring a new product to market, PD is a necessary process that must be given top priority. Depending on your objectives, there are many methods to approach this process. We hope that this chapter has been helpful in laying out the steps involved in PD and the many strategies for making your next launch successful.

For a product to be successful, many factors need to be taken into account, and we must be very careful about what we sell to our customers.

One of a company's most crucial endeavors is the creation and development of new goods. Suppliers and sourcing specialists contribute most significantly to this process. The key contributions may be seen in the quality, price, and time to market categories.

Only in a setting of mutual cooperation between the stakeholders can the team successfully build a product. Companies must assign their sourcing specialists strategically and concurrently with the development of their goods and manufacturing procedures. Because sourcing experts identify issues early, it is crucial for businesses to incorporate them in PD, process design, and engineering activities early on.

References

1. Eliashberg, J., Lilien, G.L., and Rao, V.R., "12 Minimizing Technological Oversights: A Marketing Research Perspective," in March, J., Garud, R., Nayyar, P., and Shapira, Z. (Eds), *Technological Innovation: Oversights and Foresights* (Cambridge, UK: Cambridge University Press, 1997), 214, doi:10.1017/CBO9780511896613.014.

2. Ulrich, K.T. and Eppinger, S.D., *Product Design and Development*, 3rd ed. (New York: McGraw Hill, 2004).

3. Cooper, R.G. and Kleinschmidt, E.J., "New Products: What Separates Winners from Losers?" *Journal of Product Innovation Management* 4 (1987): 169-184.

4. Ofek, E. and Srinivasan, V., "How Much Does the Market Value an Improvement in a Product Attribute?" *Marketing Science* 21, no. 4 (2002): 398-411.

5. Schmalensee, R. and Thisse, J.-F., "Perceptual Maps and the Optimal Location of New Products: An Integrative Essay," *International Journal of Research in Marketing* 5 (1988): 225-259.

6. Carrillo, J.E. and Franza, R.M., "Investing in Product Development and Production Capabilities: The Crucial Linkage between Time-to-Market and Ramp-Up Time," *European Journal of Operational Research* 171, no. 2 (2006): 536-556.

7. Loch, C.H., Pich, M.T., Terwiesch, C., and Urbschat, M., "Selecting R&D Projects at BMW: A Case Study of Adopting Mathematical Programming Models," *IEEE Transactions on Engineering Management* 48, no. 1 (2001): 70-80.

8. González, M.O.A. and Toledo, J.C., "Customer Integration in the Pre-Development Stage of New Products: Management Process Proposal," in Lindemann, U., Srinivasan, V., Kim, Y.S., Lee, S.W. et al. (Eds), *Proceedings of the 19th International Conference on Engineering Design (ICED13) Design for Harmonies Vol. 3: Design Organisation and Management Seoul* (Scotland: Design Society, 2013).

9. Flores, F., "Innovation by Listening Carefully to Customers," *Long Range Plan* 26, no. 3 (1993): 95-102.

10. Maguire, S., Koh, S.C.L., and Huang, C., "Identifying the Range of Customer Listening Tools: A Logical Pre-Cursor to CRM?" *Ind Manage Data Syst* 107, no. 4 (2007): 567-586.

11. Xie, M., Tan, K.C., and Goh, T.N., *Advanced QFD Applications* (Wisconsin, MI: ASQ Quality Press, 2003).

12. Griffin, A. and Hauser, R.J., "The Voice of Customer," *Mark Sci* 12, no. 1 (1992): 1-27.

13. Reichwald, R., Stotko, C.M., and Pillar, F.T., "Distributed Minifactory Networks as a Form of Real-Time Enterprise: Concept, Flexibility Potential and Case Studies," in Kuhlin, B. and Thielmann, H. (Eds), *The Practical Real-Time Enterprise—Facts and Perspectives* (Berlin/Heidelberg: Springer, 2005), 403-434.

14. Reed, F.M. and Walsh, K., "Enhancing Technological Capability through Supplier Development: A Study of the U.K. Aerospace Industry," *IEEE Trans Eng Manage* 49, no. 3 (2002): 231-241.

15. Gunasekaran, N., Arunachalam, V.P., and Vinu Selva Kumar, A., "Web-Enabled Integration of the Voice-of-Customer for Continuous Improvement and Product Development," *International Journal of Services and Operations Management* 2, no. 1 (2006): 78-94.

16. Reinertsen, D.G. and Smith, P.G., "The Strategist's Role in Shortening Product Development," *The Journal of Business Strategy* 12, no. 4 (1991): 18-22.

17. Karagozoglu, N. and Brown, W.B., "Time-Based Management of the New Product Development Process," *J Prod Innov Manag* 10 (1993): 204-215.

18. Chen, J., Reilly, R.R., and Lynn, G.S., "The Impacts of Speed-to-Market on New Product Success: The Moderating Effects of Uncertainty," *IEEE Trans Eng Manage* 52, no. 2 (2005): 199-212.

19. Jetter, A.J., "Educating the Guess: Strategies, Concepts Tools for the Fuzzy Front End of Product Development," in *Proceedings of PICMET*, Portland, OR, 2003, 261-273.

20. Szakonyi, R., "Dealing with a Nonobvious Source of Problems Related to Selecting R&D to Meet Customers' Future Needs: Weakness within an R&D Organization's and within a Marketing Organization's Individual Operations," *IEEE Trans Eng Manage* 35, no. 1 (1988): 37-41.

21. McLaughlin, C.P. and Stratman, J.K., "Improving the Quality of Corporate Technical Planning: Dynamic Analogues of QFD," *R&D Management* 27, no. 3 (1997): 269-279.

22. Wu, H.-H., Liao, A.Y.H., and Wang, P.C., "Using Grey Theory in Quality Function Deployment to Analyse Dynamic Customer Requirements," *Int J Adv Manuf Tech* 25 (2005): 1241-1247.

23. Chen, C.-H. and Yan, W., "An In-Process Customer Utility Prediction System for Product Conceptualization," *Expert Systems Appl* 34, no. 4 (2008): 2555-2567.

2

Learning

Product Development Is Learning

Popularly, learning is more associated with formal institutions, while training and development are associated with learning that occurs in organizations. However, the concept of learning in organizations has only begun to enter the PD arena recently.

As opposed to training, learning is typically described more broadly as a process that includes both education and training. Learning is "a self-directed, work-based process leading to enhanced adaptive capacity," according to one definition. Learning is context based and, therefore, work environment based. Developing human resources might be thought of as its core because it is a continuous, lifelong journey which is not always well thought out or deliberate. In the automotive industry, there is a range of things to know, from product to process to industry and customer.

According to this process, learning is at work whenever "people can demonstrate that they know something that they did not know before (insights and realizations as well as facts) and when they do something they could not do before (skills)." It also includes acquiring insights and factual knowledge [1].

It is essential to incorporate learning into every aspect of our work lives. Knowledge is especially true for PD. Continued learning should not be restricted to formal institutions but ideally included as part of our daily lives. Capturing and disseminating that learning is also essential. This chapter demonstrates how PD is closely related to and associated with learning (Figure 2.1).

FIGURE 2.1   PD for vehicles has been going on forever.

What Is Learning?

In our experience and a substantial portion of this book, communication is critical to learning, PD, project and product management, and any organization's efforts.

Words have meanings and misunderstandings or misconstrued communication; one with two or more interpretations ascribed to the same term can result in drastically or subtly different outcomes. We may never know the difference until we see the impact. Therefore, clear communication is essential for PD and business. In addition, clear communication requires the ability to ask questions.

One of us worked as a global company's verification and test group manager. There was another part of the company that developed customer-specific vehicle solutions. The organization wanted to make this customer-specific solution available to all vehicle models, not just customer specific. Having witnessed many failures because of shortcutting a customer adaption to a standard offering, we asked questions about this proposed order book update. For example,

- What is the volume of customer adaptations annually?
- What are the failures in the field?
- How many and costly are these events?

These are reasonable questions to ask to determine the best course of action. Unfortunately, these questions prompted an executive to accuse us of being obstructionist. There are two things we learn from this scenario.

Culture of Questions

An accusatory tone is not the attitude or response that facilitates open dialogue and questioning. As it turns out, there were many field failures for this adaptation, a costly quality issue for the company. Ultimately, the customer adaptation was not moved to the order book, avoiding enlarging the poor-quality financial exposure. This attack on the team member's objective for asking questions does not promote questioning. Questions are one of the ways we learn and seek to understand.

Learn from Failure

The result of the engineering effort that went into the customer-specific design was readily available. The organization's systems gathered and retained field data that would present enough information for anybody skilled to assess the product capability in the field. If the organization wanted to know the capability or quality of the developed part in the customer vehicle, all one needed to do was look.

Common Lexicon

Effective PD relies upon constant and clear communication. We have a project manager friend that once said, "if we did everything correctly, we would not require testing." He is partially correct. The problem is that there are thousands and thousands of interactions and interpretations of verbal and written information.

One of us believes so strongly in a common lexicon that at the beginning of creating a new verification and test group, one of the objectives was to take the entire team (in parts) to sit through software testing certification training. Not that we required the team members to obtain the certification, but because this created a baseline of understanding we could adapt and build to meet our specific needs.

The ability to learn from each other requires the ability to share information. A significant part of our learning will develop a common but not coercive understanding of the processes and words. We have worked in companies that had many TLA (three-letter acronyms) and sometimes had associated definitions. Consider basic comprehension, concepts, and phrases associated with PD. For example, is there a difference between verification and validation (V&V)? Are there internal or external standards we should use? How does this change with a change in the industry? A uniform terminology guarantees that the concepts are communicated and understood. For example, we wrote our first book *Project Management of Complex and Embedded Systems* because we heard project managers on a project use concepts errantly.

For instance, according to Merriam-Webster [2], the act or experience of one who learns is called "learning," it also refers to the acquisition of information or skills via teaching or study as well as the alteration of behavioral tendencies through exposure.

This definition argues that there are many ways in which we might learn. Additionally, how we learn will influence our behavior, depending on the degree to which we are receptive to new experiences, and this process is ongoing.

Functionally, learning has been defined as changes in behavior brought on by experience, and mechanistically, learning has been characterized as changes in the organism brought on by experience.

Problematic definitions exist for both categories. We define learning as ontogenetic adaptation, or modifications in an organism's behavior brought on by patterns in its environment. This functional definition not only addresses the shortcomings of existing definitions but also offers significant benefits for the study of cognitive learning [3].

We are all likely gaining new knowledge. Likely, we are all aware that the ability to put what we have learned into practice truly determines how well we have learned. Moreover, a positive outcome of applying that learning may prove that we have learned something about the context for using that learning.

NPD consists of many domains; becoming an expert requires time and considerable effort. We are lucky if we can master a few of these. To learn as much as possible to develop a product, one of us obtained a Master's in Business Administration in Marketing and a Master's in Project Management. It is not sufficient that the team has excellent engineering and innovation. We must understand the customer and manage the project effectively to produce the desired results.

Quotes on Learning and Not Learning

Each project is different and has its challenges, but these challenges may have a common theme that can apply to other projects. For example, from experience, schedule challenges are a common theme. We can see how this repeated failure to learn costs our company much money in projects, functional areas, and business processes. The ability to accommodate growth by learning and changing characterizes organizations with good project management. It is okay to make mistakes because that is how we learn.

> Good judgment comes from experience, and experience comes from bad judgment. —Rita Mae Brown

On the other hand, we should not routinely burn our hands on the same stovetop and then appear startled when it does. You have an issue with learning if you notice that your project or organization repeats the same set of mistakes repeatedly. It is impossible to know everything, but if you keep your mind open to learning, you will continually expand your knowledge. From experience, a recurring failure mode is not a motivating factor for the team members. We heard a manager say once, "it is okay to make mistakes, but can't we make different mistakes?"

> There is only one thing more painful than learning from experience, and that is not learning from experience.—Laurence J. Peter

It is possible that, at times, it will appear as though the company is incapable of learning. Learning at an organization's level is like balancing on a tightrope. Learn something valuable and essential, but do not forget to have an open mind about other possibilities that could be beneficial in the future. To do this, you need to have a firm grasp on what is truly significant and what only appears to be significant.

> We should be careful to get out of an experience all the wisdom that is in it–not like the cat that sits on a hot stove lid. She will never sit down on a hot lid again–and that is well; but also she will never sit down on a cold one anymore.—Mark Twain

Bloom's Taxonomy (Depth and Breadth of Learning)

Bloom's taxonomy was developed to discuss learning objectives and exchange assessment tools among teachers using a common language (Figure 2.2). It provides a hierarchical scale to define the amount of skill necessary to produce each quantifiable student result.

FIGURE 2.2 Knowledge is not a discrete have or have not; Bloom's taxonomy demonstrates an example of this gradiation.

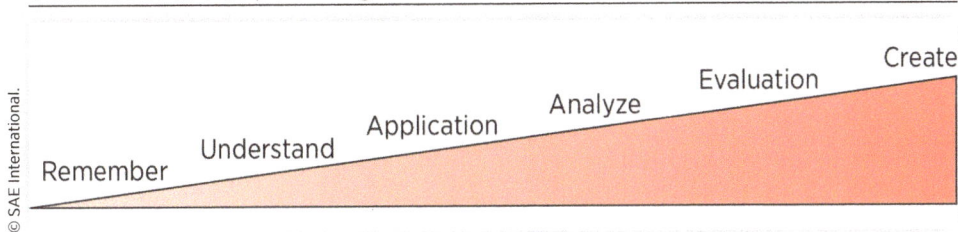

© SAE International.

Initially, three different taxonomies depend on the desired target level of knowledge. They include psychomotor (skills based) goals, cognitive (knowledge based) goals, and, lastly, affective (heart-based feelings) goals.

The cognitive taxonomy is the most prevalent, adaptable, and arguably best suited for learning as it applies to PD. A revised Bloom's taxonomy has six levels. These include Remembering, Understanding, Application, Analyzing, Evaluating, Creating. Each level increases in understanding and the individual's ability to apply. Bloom's taxonomy applied to PD is found below. Each subsequent level represents a greater degree of competence and capability (Figure 2.3).

FIGURE 2.3 PD starts small where we learn in the service of the big objective.

- **Remembering:** The individual or team can define concepts related to PD via recall.

- **Understanding:** The individual or team can follow the rationale for the specific process or knowledge area.

- **Application:** The individual or team can apply the principles in a specific context to a PD effort. Application is just that, using what is known.

- **Analyzing:** The individual or team can compare a situation to past events or the ideal approach. The team member can compare and contrast the situation against standards.

- **Evaluating:** The individual or team can determine discrepancies and which standard or approach may best apply to the situation.

- **Creating:** The individual or team can synthesize multiple approaches into a specific ideal system to particular circumstances. To get to this point, the team member will have gone through each of the preceding via experiential learning.

Benefit of Learning Fast

A learning organization promotes and fosters learning at all levels for it to alter and adapt to the ever-evolving and ever-changing work environment [4].

Learning fast as individuals, teams, and an organization is essential in PD. Some of the benefits of learning fast in the PD process include

1. Getting ahead of the competition
   Learning and disseminating that learning fast in PD is a competitive advantage. Specifically, acquiring and spreading information more quickly help reduce market time.

2. Better and faster decision-making abilities
   Better decision-making skills include ruling out the unwanted from the desired a separating the wheat from the chaff, as it were. It is knowing what is essential and providing focus.

3. Morale or esprit de corps
   We hit on this a bit earlier; making the same mistake over and over does not facilitate motivation. Spending time cleaning up the results of poor decisions, for example, programming vehicles in the rainy and cold mountain air, over and over again. Not much new learning. Team members working over the holiday season to meet project objectives because of poor decisions elsewhere in the organization? Again.

Individual Learning

Individual learning is a set of clear ideas and exercises that allow someone to learn, develop a sense of identity by creating their vision, and have an objective perspective on the world.

1. Profit on employees' strengths
   Individual learning helps organizations improve organization. The individual is the basic unit of knowledge and talent of the organization. As a result, employees choosing subjects/areas that pique their interests are positioned to perform their best. It is like having experts in different fields on the team! With a little more effort, these individuals can be coaches in their areas of expertise.

2. Less wastage of training time
   Individual learning in organizations reduces the number of redundant lessons and time wastage. In addition, experience suggests that knowledge gained and applied on the job benefits all. Education is context based, and the best way to get that, in our opinion, is in the actual work context. Therefore, learning on the job can be more beneficial than formal or outside training. It is just in time consistent with the seven wastes.

3. Setting personal visions
   Individual learning helps employees to discover their strengths and interests. This information knowledge allows them to set individual goals and career paths based on their interests and personal objectives.

   The consequence is increased productivity in the workplace.

Team Learning

Team learning is considered the process of aligning and building the skills of a team to get the results its members want [4]. The sum or amalgam of individual learning coupled with a shared vision is essential for team and organization growth. Learning and shared vision are part of the road to creating a team out of a collection of individuals. Peter Senge says that when teams learn together, not only can the organization benefit but the team members will also grow faster than they would have otherwise.

The foundation of team learning is a "dialogue," or the team members' capacity to set aside their viewpoints and begin "thinking together." The Greeks understood dia-logos as a free exchange of meaning among participants that enabled the group to learn things no individual could have known. It also requires learning to spot team behavior patterns that make it hard to understand.

Benefits of Team Learning in an Organization

1. Improves productivity
   Collaborative learning of new concepts and techniques provides employees with a shared vocabulary and knowledge that enables them to create more fruitful and meaningful goals for the business unit. An environment open to exploration helps the team to find other ways of completing tasks more quickly or adapt an existing process to meet the circumstances.

2. Encourages collaboration
   It is well known that human experience has a proven ability to bond people together in remarkable ways. From experience, the trials the team encounters are opportunities for moving our collection of individuals to a team. When employees learn together, they can share the experience of group discussions and develop cognitive frameworks for assessing and applying methodologies and theories to their objectives and tasks. As a result, they create a togetherness that will enhance their capacity to work productively together.

3. Help builds employee relationships
   Cohort-based learning is a method of learning involving group learning. Challenges presented to the team offer on-the-job cohort learning that helps employees learn together. This learning might be a little tumultuous as the team attempts to glean the truth. However, this can improve team interactions over time, provided we constructively handle disagreements and conflict.

   The cohort approach to education which allows students to move through the curriculum at the same rate gained popularity in the 1990s. This improvement was after it was discovered that students could motivate each other to do better, which increased the number of students who stayed in school and finished their courses.

4. Improves employee engagement
   When you train a team at the same time, you hit a lot of the things that make people more interested. These include ways for employees to grow and have their voices heard.

   Teamwork teaches employees how to share their thoughts and opinions in a group environment in a way that brings out individual confidence and in a courteous manner.

   Even though engagement and job satisfaction are not the same, employees who are engaged tend to be much happier at work. They find meaning and purpose in what they do, and when they invest in their work products, they feel like they have reached their goals (Figure 2.4).

FIGURE 2.4 In general, there are two camps for PD, agile and waterfall or stage gate.

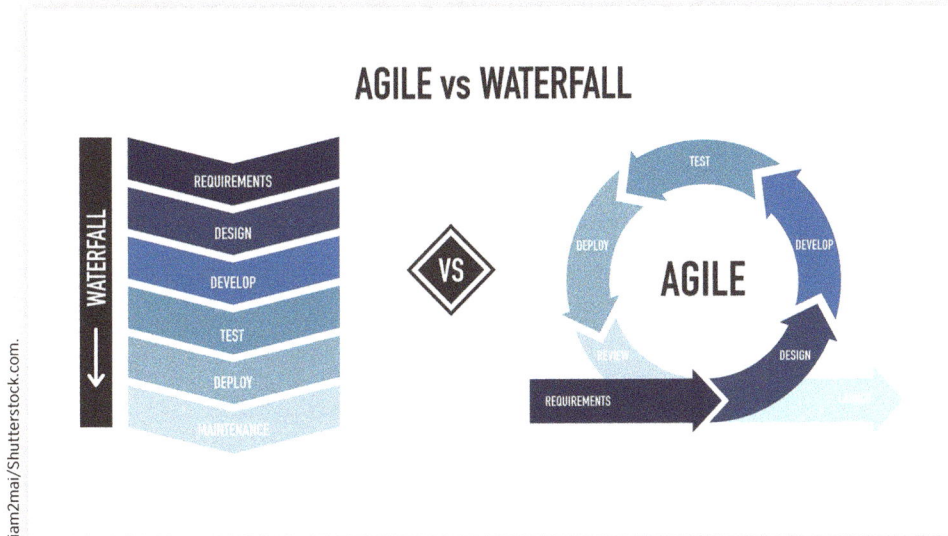

Agile Overview

Agile methods of working have gained traction over the years, having originated as a way of software development. However, this approach is not just applicable to software creation; but is used in other contexts. For example, we have adopted elements of scrum in a verification and test group amid a stage gate project.

Agile puts its emphasis on the job itself rather than on the processes involved in its completion. Moreover, it attempts to create an atmosphere that is amenable to transforming a collection of individuals into not just a team but a self-directed and self-organizing team.

An agile approach takes a different approach. Instead of exercising control over the work from the top down, the company runs its business in a way that makes it easier for people to learn and encourages the growth of teams into self-directed work units.

There is an agile saying that goes: *fail fast and fail often*. This proverb inspires a team to take calculated risks and respond to setbacks differently. In particular, there should be no fear of failing, and when it does occur, we do not hide the result; instead, it should be explored and disseminated so that everyone may learn from it.

Learning Environment

The learning environment is a term that refers to the physical location, means of teaching, contexts, mode of learning, and learning resources that one uses to acquire knowledge. A conducive learning environment is essential for a successful learning experience. Therefore, in PD, organizations should make a point of providing a conducive learning environment for the team.

Benefits of Controlled Failure

Failure is like a teacher that helps individuals learn how to move forward. Businesses that do not learn will have recurring failures and incur customer and monetary losses as a consequence of the failure. Some of these failures may end up driving the company out of business or remain in business but significantly reduced in capability via loss of income and talent.

In case you are wondering, those that keep an organization in business are better. Failure is part of life and it is inevitable. However, there is a way to fail, just enough to get back up and learn a lesson or two, and it is called controlled failure. Controlled failure is suitable for businesses because it helps them know and understand their business and what works for their market and what does not.

Why Push to the Edge Excluding Previously Thought to Be Constraints?

Planning to learn from an activity (experience) needs to be built into as many activities as possible to get away from the "that is how we have always done it" state of mind.

Additionally, any continuous improvement initiative of the organization will require a constant and consistent application of effort into improvement. Additionally, when you plan some form of improvement from an activity, it inherently implies that a deeper understanding of the activity exists and is beneficial to acquire.

While this might not be the case, it will become evident when we assess what is learned is contrary to what was anticipated to be discovered. Either way, the planning and the activity will produce some form of knowledge via individual experiences.

Whether we are after a better system understanding, an improvement in an activity, or, even better, both, this is a prime example of a learning organization: planning to learn.

While many organizations are reactionary, using cause mapping and critiques to hopefully correct an issue, a learning organization seeks opportunities to learn from activities rather than waiting for issues to arise.

However, not even the best learning organization is immune from an unforeseen development (specific risk), but the more we know, the easier it is to identify risks to the project and the organization.

While a non-learning organization commonly seeks a rapid restoration to business as normal, not exploring the root cause (or poorly executed root cause), a learning organization will pursue the true underlying issue to allow a more effective solution.

These will ideally culminate in checkpoints for validation to ensure those things learned are applied and produce the desired project effectiveness or need for modification.

Learning by Discovering Boundaries

One of the reasons why statistical analyses are essential in NPD is because they help to distinguish between nonrandom variation (controllable variation) and random variation. Controlling random variation is notoriously difficult.

Even so, we have seen instances when variance could be reduced through the careful application of specified trials and experimentation. Moving the mean is a significantly more prevalent practice among practitioners than attempting to "fix" the variance.

It is important for project managers to understand which factors are controllable and which ones lay outside their scope. Once the manifestation of this awareness occurs, project managers are able to proceed smoothly because they understand which variance are controllable and which ones are not.

Moreover, the charts have the potential to offer management a helpful visual signal of what is actually going on while the process is being carried out. Finding project material that can be represented using control charts is the aspect of the statistics and control chart method that presents the most challenge.

According to our observations, project managers frequently approach each new endeavor as if it were so singular that no lessons could possibly be learned from previous endeavors.

In this line of business, we have witnessed managers use hope as their strategy for product innovation and in the management of projects in general. We give it this name, hope, because when we make decisions, we do so with the intention of bringing about a particular outcome, even if we may not have a good understanding of what it takes to be successful.

Even when we are in possession of this "knowledge," our organization may choose to disregard what is being communicated to accomplish this goal. We are not talking about an isolated incident in which something slips through the cracks here and there.

We are discussing the instances when people around us have been taking measures to have a better understanding of the challenges faced by that business as well as the opportunities it presents. The individual who is proving us wrong with this information is swiftly dismissed as not being a team member or as merely a naysayer with a negative attitude. This is done rather than building upon what has already been established.

An engineer who works in production will pay close attention to the capacities of the machinery they use. If, for instance, the required or desired product throughput via that piece of equipment is more than the amount of product that the equipment is capable of generating, then an alternative plan of action will be necessary.

It is feasible to make adjustments to the machine, which would involve personalizing the equipment to achieve better results. It is possible that otherwise upgrading the equipment would be judged required.

Consider investing in an additional piece of machinery to cater to the increased need for throughput. Investing in an additional piece of machinery might solve this problem; then, with these two machines working together, the throughput requirements could be satisfied.

If a piece of machinery has a throughput constraint, a sensible engineer will not likely try to force the product that is being demanded through the machinery and then expect everything to go smoothly.

Consider an item that is crafted from a certain kind of plastic that has a predetermined temperature at which it melts. Above a certain level of heat stimuli, it is quite unlikely that this product will keep the features that the customer wants or fulfill the requirements that they have specified.

It is not plausible that a reasonable individual would violate the physical qualities of the material and then complain about the failure that was expected to occur. Before employing the plastic in the design of the product, the person creating the product would benefit immensely from having a thorough understanding of the attribute of the plastic.

What exactly links all these things together? To tell you the truth, this is the kind of stuff that takes place daily in businesses! This happens in PD and project management.

Some people think that "giving it our best shot" will be enough to save us. Give it the proverbial old college try. Those persons who point to genuine data to highlight the "reality" can be ridiculed for being pessimists, and they are not considered to be team members—from experience.

These individuals have an unhealthy level of pessimism. What kind of message does it send to the members of the team who choose to disregard the past?

This age-old proverb is commonly attributed to George Santayana and states: "Those who cannot remember the past are condemned to repeat it."

In Webster's definition of learning, we see the statement, "modification of behavioral tendency by experience." There is an adage that experience is the best teacher, but do we behave in a project or as an organization in conformance to this principle?

Everything that we do creates a situation wherein someone else can experience something and therefore learn. Is the experience we present others promoting one experience (lesson) and expecting a different result (learning result)?

As we have briefly touched on in this chapter and will additionally in the later chapters, experience plays an instrumental role in both motivation and development. This is not to say that we should always provide people with what would be considered a positive experience; we should prove or facilitate the experience that is related to the modification of behavior that is desired.

On its surface, this sounds like manipulation, but if this is done with an open dialogue between the parties involved, it is not manipulation, it is mentoring. Mentoring is thought of as from a senior individual to a more junior one, but when we apply how experience teaches it must provide all people involved with some form of development (Figure 2.5).

FIGURE 2.5   Assumptions, left unvetted, can cause ruin in our development effort.

Kim Britten/Shutterstock.com.

What We Do Not Know and Assumptions

For the mere reason that the experience of failure is not pleasant to anybody, most people fail to know that one can leverage failure to their advantage and benefit from it.

One of the things we do not know about failure is that it contributes to later success. How? Northwestern University's Kellogg School of Management did a study on students in relation to securing funding for their research.

This study found that the students who were not able to secure funding earlier on had a greater chance of success than those who did. Those who were not able to secure funds early have a 6.1% more of a chance to publish a better paper than those who did.

Second, failure makes and increases a person's resilience. Going through failure and being knocked down is as unpleasant as it sounds.

However, through getting back up, we build such an undefeated resilience and not to mention the lessons and tips we gain from such an experience. What we get from being knocked down and getting back up is that we have the strength to survive such a hit.

One of the biggest assumptions about controlled failure is that it is bad. Just to make things clear, failure is not a bad thing, but the experience of failure is very unpleasant.

With controlled failure, it is technically deliberately failing to learn what works. One thing about controlled failure is that the failure might not necessarily happen, and when it does, it is calculated for; hence, the damages may not be very severe (Figure 2.6).

FIGURE 2.6 With reward comes risk, our efforts balance the acceptable risk for a desired benefit.

Olivier Le Moal/Shutterstock.com.

Calculated Risk

Calculated risk in PD refers to a project or the actual or potential outcome of a project whose likelihood of failure has been predicted. Risk-taking in entrepreneurship is beneficial for businesses. More so, calculated risks have a better chance for future reward (Figure 2.7).

Whale Design/Shutterstock.com.

FIGURE 2.7 Ideally, the work itself offers an opportunity for individual, team, and organizational learning.

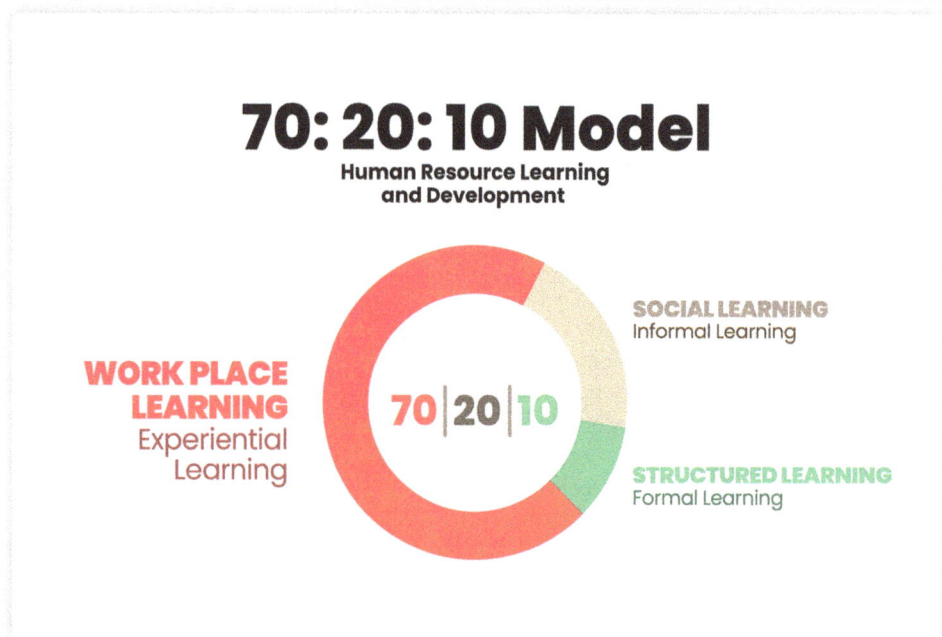

Experiential Learning

Experiential learning is the kind of learning whereby learning takes place by the process of doing. Or in other words, it is the process of learning through experience.

Below are some of the elements involved in experiential learning:

- A planned learning experience that offers the chance to gain knowledge through unavoidable outcomes, failures, and accomplishments.

- Critical analysis, reflection, and synthesis.

- The developmental team is given the opportunity to take initiative, participate in decision-making, and accept responsibility for the outcomes of their actions.

Experiential learning is important for PD because, like the famous saying goes, experience is the best teacher. The act of reflecting on what one has learned both during and after one's experiences is an essential part of the process of learning. Because of this, introspection, analysis, synthesis, and critical thinking are produced [5, 6].

Fail Fast

Despite being widely associated with the industrial mentality of Silicon Valley, the expression "fail quickly" is not native to this area. The phrase "intelligent rapid failure" was first used in 1989 by Jack V. Matson, a professor of environmental engineering at Pennsylvania State University. When a risk is taken intelligently, fewer resources (such as time and money) are lost if it fails, according to Matson's definition of "intelligent" in his 1991 book *The Art of Innovation: Using Intelligent Fast Failure*: "When you take a risk, you want to learn as much as you can about what happened and why through obtaining feedback." For you to quickly ascertain what happened, "fast" suggests a hurried risk. Failure shows that the bulk of plans won't work out as expected. Even though the majority will fail, it is in the process of failing that you will uncover the misrepresented information that will enable you to take smart risks [7].

It is hardly a big surprise that business leaders have to deal with more and more uncertainty at work with the changing environment. It should come as no surprise either that under uncertain circumstances, failures outnumber successes.

Yet, for some inexplicable reason, we do not create companies in a way that they can mitigate, manage, and learn from failures.

Fail fast and fail often. The agile saying discussed earlier encourages viewing failure as an opportunity to learn instead of a knockdown. In PD and in life as a whole, failure occurs a lot of times.

A common saying by Mickey Rooney that says "You always pass failure on your way to success" emphasizes the importance of failure in life because without failure, there is most likely no success.

Through failing is how you know what not to do next time, and that is how you get it right at last. The same applies to PD; feasibility is discovered through trial and error.

What puts winners and successful businesses at the top is their ability to learn from their failures. A change of perspective is essential when things go wrong because it is a form of motivation.

Motivation acts as a fuel to keep going. In the context of agile and based on past experience, "fail fast" refers to the practice of designing work-based experiments from which to draw conclusions and gain knowledge. To acquire knowledge in a timely manner, the duration of these trials have to be kept brief.

For instance, if there is a concept for a certain design approach but an organization is unsure whether or not it will work. They should think up a little experiment that revolves around that design approach and determine what measurements will allow them to forecast whether or not this approach will be successful.

The idea behind the phrase "fail often" is that we ought to be continuously learning and pushing the boundaries of our capabilities, but in doing so, we expose ourselves to the possibility of experiencing a string of failures. For the sake of clarity, the point is not to fail simply for the purpose of failing.

Failure should be tolerated in the workplace, and we should not let the fear of it prevent us from attempting new things or expanding our knowledge. The workplace atmosphere should be set up in such a way that failure is acceptable.

All these things require a setting in which it is accepted that a person may make mistakes in their line of work and in their career, and in which the members of the team are always looking for new and improved ways to carry out the tasks at hand. However, taking the proverb less seriously and interpreting it more flippantly could have some unforeseen repercussions.

> When leaders do not fully comprehend or appreciate a term, the result can be the exact reverse of what they intend. Worse, when we muddy the waters with words such as "fail fast, fail often," it can create irreversible harm, particularly to company culture.—Forbes [8]

Time to Market

According to earlier studies, experience timing plays a crucial role in learning when it comes to PD. If a business is participating in PD, the worst thing it can do for itself is to fail slowly.

Failures offer businesses insight into what might be going wrong and serve as feedback that might enhance performance moving ahead. The timing of the failure is crucial to the learning outcomes for the companies since it affects how quickly the companies receive feedback on a project.

Failing fast and failing often is beneficial for a business in the long run because they can switch up quickly and change course. According to experimental findings, participants will make fewer mistakes and learn more rapidly when they receive prompt feedback.

Decision-makers may find it challenging to analyze links between actions and results when they receive delayed data because it may be muddied with noise. The value of early feedback and its beneficial effects on performance are emphasized.

Yet, time and again, people who have ideas waste too much time fixating on the big idea or the theory behind the idea. However, at the end of the day, what ultimately matters the most is how fast and effectively you can reach a large enough percentage of your target market. Or else, your idea faces a risk of ending up dead in the water.

It is important to prioritize speed while launching a product or service. Treat your idea as a popularity contest, the faster it is discovered, and by a lot of people, the better chances it has for success. Waiting in some cases could mean more harm for a product than the opposite. It is better to try and fail than not try at all. I bet it would be painful to see your idea flourishing when it was brought to market by someone else.

Why Is It Important to Learn about Failures Early?

Small failures may produce the most learning when there is prompt feedback, allowing the company to experiment with different approaches and learn as it goes.

For instance, a chemical may fail in the pharmaceutical industry either very early on, even before preclinical testing, or many years later, during late-stage clinical trials. The external environment and internal procedures both play a role in determining whether a company would experience failure relatively early or late in the development of an innovation.

The likelihood of discovering problems early and learning from them will be higher than it would be if internal processes inside the company were not specifically focused on this process. Sometimes businesses are forced to wait before judging a technology to be a success or failure because they are so confined by the signals from the outside environment.

Interim feedback encourages businesses to hunt for solutions by assisting them in identifying potential difficulties. This view is supported by anecdotal information about the strategy of highly innovative businesses.

Early input during the R&D process enables businesses to manage resources and restrict resource investment to fruitless areas. In contrast to failures that occur later in the R&D process, early failures enable businesses to experiment in more ways.

Contrarily, when feedback is received later in the R&D process, it may be difficult to identify the precise choices or behaviors that contributed to the failure, confusing learning.

Additionally, subsequent failures may escalate commitment and force a company to continue making associated expenditures.

R&D in high-technology industries is path dependent, and it is challenging for businesses to shift course after making sizable progress. Early feedback on technology may make it simpler for a company to restructure its R&D spending and better use the lessons learned from mistakes [9].

Failing Is Learning

Failure is a necessary component of learning, which is why the proverb "If, at first, you don't succeed, try, try again" was first used to motivate kids to keep working hard in class.

Theoretically, we recognize that failure is a part of the scientific process in the same way that we anticipate people to fail as they learn new abilities. The easiest method to reject a hypothesis is for an experiment that tests the hypothesis to not provide the predicted outcome.

Therefore, in theory, academia should be open to failure because it is a community based on the twin pillars of scientifically based instruction and research, both of which depend heavily on failure.

However, those who pursue jobs in research are frequently those who are accustomed to being the "smart" ones and who did not encounter many failures in school. Admitting to less-than-ideal results is difficult in a competitive climate when people attempt to prove their qualifications to advance in (or keep!) their work.

Additionally, failure becomes more stigmatized the less we talk about it. In international development, the inability to discuss failure covers a wide range of sins.

We frequently encounter news articles, eye-catching films, and academic papers describing the early stages of the development of technology that aim to address issues in underdeveloped nations. When tests begin to reveal that findings are not as favorable as initial assumptions, far too many of these quietly disappear.

Instead of showing off just the successful parts of the process, more celebrities should encourage people to take failure as a positive thing and not give up. Clearly, everyone fails at some point, and the world would be a better place if failure was less stigmatized.

How this Works

Failure to exploit and learn from failure arises automatically whenever there are no plans for learning from failure. One must be willing to take a lesson or two from failure.

The overemphasis on the bottom line has consequences for actions, and these consequences can restrict some of the actions that are considered suitable, legal, or even wise.

There are times when there are opportunities for the team to learn, but instead of taking that learning potential and the risk, a safe strategy is selected, which fits within what is already known by the team. This is referred to as a failure to exploit.

This practically amounts to missing out on the chance to increase one's level of knowledge, which is a form of learning. Within the agile community, which is a method of managing software development projects.

There are many different interpretations of this proverb, but the central idea is that nothing new can be learned without taking some risk and that we should not be afraid of failing to learn new things. Instead, we should take risks that are beneficial to our learning and not let the possibility of failing prevent us from doing so. The culture of the company effects this mindset of continuous learning.

Delay Deciding

It is economically beneficial to postpone irreversible actions or decisions until the uncertainty is lessened. It leads to better judgments, minimizes risk, aids in the management of complexity, reduces waste, and keeps consumers happy.

Delaying decisions, on the other hand, often comes at a cost. The term delay deciding in the PD process means the part of the process where an organization slows down on a certain decision to keep their options open and to remain at status quo.

Every person changes their mind occasionally, it is not any different for organizations. At times making rushed decisions could lead to more damage than good. Therefore, it is best to reduce the haste in decision-making from time to time.

When Is a Preannounced New Product Likely to Be Delayed?

During a new project for PD, it is not uncommon for the initiative to experience a delay in the final stages of the process or the stages that follow in the commercialization process. The primary reasons for the delay are typically discovered in the early phases of what is frequently referred to as the front end of NPD.

This phase occurs when a new product idea is first considered until the decision to begin or discontinue the PD. The front end of NPD is usually dynamic and participative, according to earlier studies on NPD management. What defines this stage are factors like high levels of complexity and ambiguity, which are brought on by the complex processing of information, competitor organizations' pressures, and improvised decision-making.

These challenging features typically result in errors, delays, and subpar goods. Prior studies have also demonstrated that the success of a product is significantly influenced by one's capacity to manage the front end of NPD when thorough product descriptions are established. If the product definitions are vague or faulty, the development process could be overly expensive or even fail. Most studies on the early stages of NPD conclude that from a managerial standpoint, this stage is different from the later stages of PD.

As a result, management of the front end of NPD should be done from a specific logic. Although the components of success in front-end management are missing from the available literature, that basically depends on anecdotal data. This means that the literature available does not have a comprehensive conceptual framework, which helps to describe, synthesize, and identify vital success factors responsible for the success of front-end management (Figure 2.8) [10].

FIGURE 2.8 Selecting the only option is not a choice and may not be the best alternative.

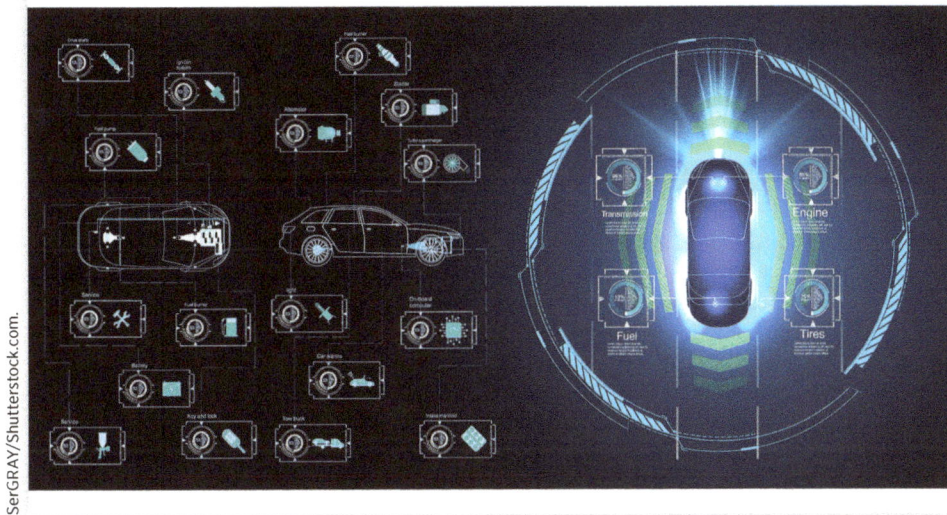

SerGRAY/Shutterstock.com.

Exploring Multiple Options

In a complicated dynamic market, when a system that prespecifies alternatives is pitted against another, the system that keeps options open prevails.

Agile techniques can be viewed as processes that provide possibilities that allow decisions to be postponed until customer needs are better understood and growing technology has had time to mature.

This does not imply that agile methods lack planning. Plans aid in clarifying perplexing circumstances, allow for the assessment of compromises, and develop patterns suitable for swift response. Therefore, plans tend to increase the ability to adapt to change. Nevertheless, a plan should not foresee specific activities based on assumptions.

Agile approaches rely on speculation, experimentation, and learning to reduce uncertainty and adjust the plan to the actual environment [11]. Options thinking is a significant tool in PD, as long as it is accompanied by the acknowledgment that alternatives are not without cost and that it takes experience and knowledge to decide which options should be kept open and explored.

Options do not ensure success; instead, they create the framework for achievement if the unpredictable future unfolds favorably. Options enable judgments based on factual information rather than conjecture.

Delay Committing to Exploring

The markets for investment products and commodities have evolved into a system known as options, which enables participants to postpone making decisions. The option allows an organization the right to decide to do something in the future, without being obligated to do it.

It is like a warranty that guarantees customer satisfaction. In this case, the organization can exercise the option if things end up going as planned, but if not, then, as stated earlier, they are not obligated to do so.

There are two directions in which uncertainty might move: unexpectedly pleasant things can happen just as easily as unexpectedly bad things can happen. Additionally, during the delay, it is possible to find better options or solutions that would be more beneficial for an organization than the previous.

A paper was published in 1988 by Harold Thimbleby on the Institute of Electrical and Electronics Engineers (IEEE) Software titled "Delaying Commitment" [12]. He observes that knowledgeable people with expertise in project management when confronted with a novel circumstance will delay making definitive decisions to conduct additional research on the matter. They do this because they are aware that postponing commitments frequently results in the discovery of novel insights. On the other hand, amateurs tend to make hasty decisions, frequently incorrect ones, since they are so focused on getting everything just right.

Once these early decisions have been made, further decisions will be built on them, making it extremely challenging to alter them later on. According to Thimbleby, early design commitment is a design failure mode that inhibits learning, heightens the impact of errors, reduces the product utility, and raises the cost of making changes.

Therefore, delay committing provides the organization with the opportunity to cash in on favorable occurrences in the future while simultaneously reducing its exposure to unfavorable occurrences. Options not only provide the opportunity to make decisions in the future but also serve as an insurance policy in case things do not go as planned (Figure 2.9).

FIGURE 2.9   We learn by experimenting: try options and learn things.

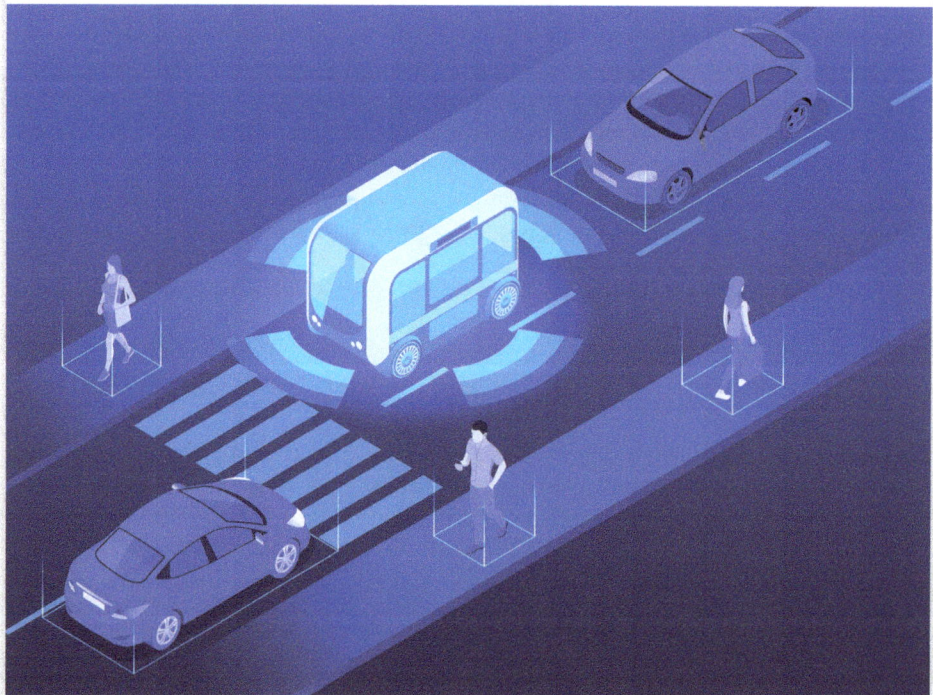

Golden Sikorka/Shutterstock.com.

Exploration and Experimentation

The consequences of a project failing late in the PD process are extremely devastating. When this happens, the amount of investment lost is huge; we are talking about opportunity costs, time, and money.

If companies took time to do experiments and explore on through which products are winners and which do not suit the category, these significant losses could be prevented. It is possible to tell which products show promise in the early stages of the development process through experimentation.

Because testing is frequently seen as a component of downstream verification, rather than as a possibility for learning during early development, businesses in many industries regrettably frequently test too little and too late. Most companies underestimate how much they can save on investments when they do the tests earlier on in the development process. As a result, they end up spending a lot of time and money fixing issues that arise later in the development stages.

What Do We Know?

Risk is marked as the simplest form of uncertainty. The possibility of different outcomes to come out of an activity is scary, and not many people or businesses like uncertainty.

Experimentation as well as exploration in PD takes time and poses a risk because it does not guarantee success. Therefore, considering the time to market factor, most organizations prefer to skip the part of the process where experimentation happens or speed up the process. Both strategies are risky, but, at times, people prefer to lean toward what feels most familiar to them. This is how biases toward one strategy over the other are created. In most of these cases the biases do not favor experimentation and exploration strategy.

Aside from experimentation and exploration being time consuming, we also tend to believe in some cases, that it does not provide an actual explanation to a solution and one can miss out on the lesson.

Unrealistic situations from experimentation and research add weight to the biasness. It is possible for data to be contaminated or wrong while still appearing to be accurate since the variables of a given product, theory, or idea are subject to such strict constraints.

The researcher may experience this in two detrimental ways. One, the data can be skewed toward a positive or desired outcome by first manipulating the variables. Second, while it may appear that the data are favorable, the positive outcomes could never be attained outside of experimental study since the real-life setting is so dissimilar from the controlled one. These and other biasness limit a lot of organizations from taking the experimentation and exploration route.

How Do We Find Out?

First, by letting go of our biases, we are able to let in contradicting information and learn something from it without invalidating it. Gaining more information broadens our scope of knowledge and experiences and therefore helps us to make more informed decisions.

Biases pose a threat to learning. Having a biased perception prevents the intake of new information that is different from what we believe to be true (Figure 2.10). However, to learn we must be willing to be proven wrong and change our perspectives. Otherwise, we may not be able to keep up with the changing trends, especially in businesses.

FIGURE 2.10 It may sound simple, but we each have experiences that color our views, and such is bias.

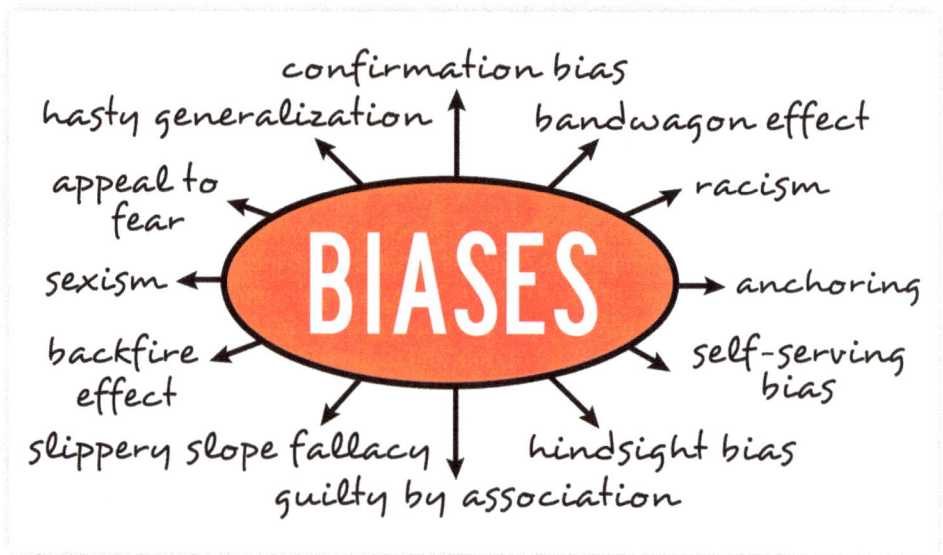

Biases

The act of being biased is the practice of favoring or opposing a specific person or entity through the use of unethical means. Biases cloud judgment and they play a role in influencing a person's decisions based on their opinions.

Confirmation Bias

Confirmation bias is the kind of bias where an individual or an organization tends to put more weight and lean toward things that confirm the beliefs they already hold. Confirmation bias can be seen as the mother of all biases. As human beings, we are more inclined toward our already existing beliefs and preconceptions, with which we tend to disregard with anything that goes against them.

We should lean toward finding faults in our ideas and questioning them rather than finding ways to defend them blindly. It would be beneficial for us to assume that we are wrong until we have concrete evidence to the contrary so that we can combat its influence.

> The first principle is that you must not fool yourself – and you are the easiest person to fool.—Richard Feynman

Optimism

Optimism also stands in as a type of bias: The kind where we have an unrealistically optimistic view of the possibility of favorable outcomes.

A positive attitude can help, but it is not smart to let it get in the way of being able to make smart decisions (although neither of these two skills is incompatible with the other).

Thinking wishfully can be a tragic irony in the sense that it can lead to even more unfavorable results, as is the case with compulsive gambling.

To have reasons to feel optimistic, we must strive to make sensible and realistic assessments first.

Availability

Availability biasness comes in whereby the idea that springs easily into the mind is what influences one's judgment.

Your memories may seem more pertinent depending on how recent they are, how emotionally powerful they are, or how uncommon they are. Because of this, you can find yourself using them far too frequently.

For instance, when we are biased toward one strategy, we can tend to only take the negative information that easily merges with our beliefs and most people's beliefs without looking at the positive side. This is because the media tends to focus on the most horrific aspects of these things (strategy).

Instead of depending just on initial impressions and the sway of one's emotions, it is best to make an effort to gather information from a variety of sources and pertinent statistics.

Negative Bias

Negativity biasness shows up when we give the things that are negative a disproportionate amount of power over our thinking.

The anguish that comes from suffering a loss or being injured is experienced more keenly and continuously than the brief satisfaction that comes from having something pleasurable. We are hardwired for survival, and our innate reluctance to experiencing pain can cloud our perception of the realities of today's society.

This type of bias plays a big role in influencing our judgment in other ways. As a result, we end up not giving the advantages nearly enough weight. This makes sense in the context of evolution, where it makes sense for us to be extremely biased to avoid hazards.

A more objective evaluation of the situation can be achieved by making lists of the advantages and disadvantages of a situation as well as thinking about it in terms of the probability involved.

Simulation and Models

Simulations and models are important for learning in the PD process. They are helpful in the transformation of a concept into a product as well as in the enhancement of the properties of an existing product. This not only increases the company's ability to compete in the market but also decreases the amount of time it takes for new developments to make it to consumers.

Experimentation

There are things we believe to be well understood and predictable; we experiment to determine the veracity of these beliefs. As reviewed in later chapters, rather than settle on one idea, we want to explore as many practical solutions as possible. These ideas are fodder for experimentation to determine which of these things will meet the attributes and constraints. Experimentation is an excellent mechanism for team learning.

If managers expand their conception of testing beyond its traditional role of verification (just viewing), they will be able to include these activities in a more comprehensive manner as part of the experimentation strategy of their company. That is experimentation through simulations and models.

When designing new products and services, it is almost impossible for the development teams to know whether or not a given concept will perform exactly as expected. This indicates that they need to devise methods to swiftly eliminate notions that are dysfunctional while simultaneously maintaining other ideas that show promise. At the same time, defective conceptions themselves create knowledge that can be used as guidance during additional rounds of experimentation through the use of simulations and models (Figure 2.11).

FIGURE 2.11   Modern computer tools allow us to readily, and with low risk, explore design alternatives.

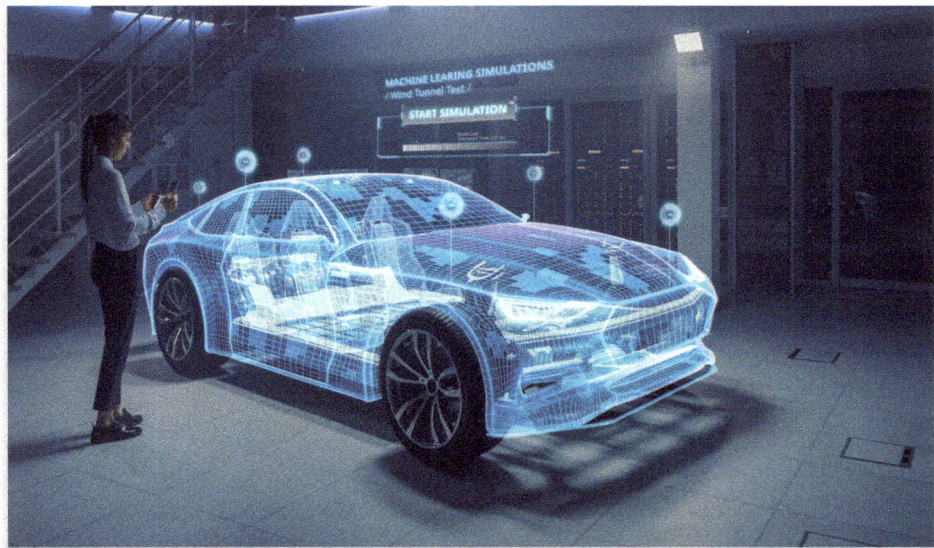

Gorodenkoff/Shutterstock.com.

Simulations

Simulations are imitations or replicas of certain things in real life. Simulations, in most cases, involve deception or illusions. In PD, simulations are the imitations of the said product that can be used to test and learn what part works and which does not work. Simulations require models to make them more realistic.

Why Simulate?

Simulations are essential in the process of PD. This is because they are used as a way of illuminating the underlying mechanics that govern the behavior of a system, which is a helpful tool to better understand how that behavior is produced.

In other words, simulations help to understand how a system works and predict how the behavior of the system might change in the future. Project managers can use simulations to escalate the system appearance and features years later, like seeing a vision.

Additionally, design simulation gives producers the opportunity to verify and confirm not just the intended function of a product that is now being developed but also the ability of the product to be manufactured. The feasibility of a product is determined through simulations.

When it comes to the process of developing new products, simulation models offer greater flexibility than their actual counterparts. It typically only takes a few clicks of a button to generate different design options, and testing those options does not require a sophisticated setup.

Because of the amount of flexibility that simulations offer, a team is able to do as many tests on a product as they possibly can without incurring many losses, which makes it easier to get a design right with the properties and all.

How Do We Know What Parameters to Include?

It is well stated above that the most important reason for performing simulations in the PD process is for the purposes of learning. The information obtained from the simulations is what organizations use to understand what works and what does not work for the product or service in question.

It is not easy to know what parameters should be or should not be included in the simulation. An organization's ability to learn from simulations, and the effectiveness of that learning, will depend on a wide variety of elements that demand commitment from both strategic and managerial personnel as well as flexibility from organizations. Therefore, there are a few factors that influence the choice of what parameters to include.

What Factors Influence the Choice of Parameters to Be Included?

1. Fidelity
   The degree to which a model accurately represents a real-world product, process, or service is what we refer to as its "fidelity." Models and prototypes which are perfect and have 100% fidelity are usually not built because an experimenter cannot capture all the qualities of a genuine situation, even if desired.

   In the early concept phase of PD, sometimes known as the "early development" phase of experimentation, it might be helpful to use lower-fidelity models if they are inexpensive and can be built rapidly. This will allow for "fast and dirty" feedback on the product.

2. Team's learning pace
   For learning to occur more efficiently, feedback should follow the action immediately [5, 13–15]. When a team receives feedback after a long time, in most cases the need for the feedback is no longer there. After waiting for a long time, the momentum is lost and people have most probably moved on to solve other problems. A company must ensure to provide feedback as fast as possible to keep the momentum running. Sometimes feedback can be provided too late for the project to go on.

3. Required resource for testing
   It is important to consider what resources are required for simulation and model creation and testing before beginning the experiments or the PD process at all. At times the ability for the provision of feedback in the PD process is highly influenced by the resources available, i.e., experts and testing equipment, among others.

4. Frequency of the experiments
   Even with the advantages of early experimentation and simulations, the question of how frequently or how many experiments should be carried out still stands. As was said earlier, the issue that plagues many businesses is that they not only test too late but also too little of their products.

The pursuit of efficacy and cost reductions frequently causes testing to be put on hold until relatively minor issues have ballooned into major catastrophes or until lost opportunities have grown into competitive dangers.

It is feasible to spend a lot of money on realistic prototype models, but the conventional wisdom says that one can cut costs and save money by putting off research and testing for as long as possible and then performing extensive "killer" tests [16].

When to Simulate

Simulations in PD should begin as early in the process as possible.

Model

Models are used as part of simulations because they bring out the attributes of a product or system as well as its current behaviors. There are four steps to creating a model:

Data

During this step, the team comes up with a design. They also come up with a clear idea of how the product is supposed to look and how it is expected to work.

Data are gathered on existing knowledge and the solutions of the model are addressed. Therefore it would make more sense if the team comes up with a list of questions, objectives, and goals that the simulated model is supposed to meet.

Building

This stage basically comprises the process of making the prototype to match the idea of its appearance as well as its features.

Compare Model Outcome to Real World

Once the building phase is over, the model is tested either in conditions that match the real world or in a laboratory. However, laboratory is not recommended because the conditions may not match the ones in the real world.

As a result, some faults may be missed in the test run or alternatively, some false mistakes may show up. In this case, the results end up needing to be more accurate.

Adapt Model to Real World

In this stage the results are analyzed and mistakes are corrected for the model to work properly in the real world. In this stage, a lot of learning occurs because the development team gets to see the promising features of the model while also eliminating the dysfunctional features.

Variables

Structured experiments demand a focused effort to manipulate or alter relevant factors. In a perfect experiment, managers or engineers would segregate an independent variable (the "cause") from a dependent variable (the "effect") and then modify the independent variable to track changes in the dependent variable.

After thorough observation and analysis, the manipulation leads to the discovery of correlations between causes and effects that, ideally, may be applied to or evaluated in different contexts.

However, things are far more complicated in the actual world. Environments are dynamic, links between variables are intricate and poorly understood, and frequently the variables themselves are ambiguous or unknowable.

Testing and prototyping can therefore be a means of providing opportunities to learn from formal and informal experimentation conducted. Traditional statistical methods and protocols enable the most effective design and analysis of experiments when all pertinent variables are known.

The statistician and geneticist Sir Ronald Aylmer Fisher introduced these methods to agricultural and biological science in the early half of the 20th century, and they are now widely utilized in many disciplines of process and product optimization [17, 18].

These structured experiments are now employed for both research where huge solution spaces are searched to identify the best response of a process and incremental process optimization [19, 20, 7].

These methods have also served as the foundation for more recent efforts to strengthen the resilience of manufacturing procedures and innovative goods [21].

However, learning through experiments is far more informal or tentative when the independent and dependent variables themselves are uncertain, unknown, or challenging to measure.

A management might be curious about how increasing an employee's incentives affects her output, while a software developer might want to know whether changing a single line of code fixes a bug. We constantly do these trial-and-error tests of this nature because they are so fundamental to the innovation process that we do them without fully realizing that they are experiments.

Furthermore, effective experimentation has ramifications for businesses in terms of how they organize, manage, and structure their innovation processes. It extends well beyond the individual or the experimental protocols. It is not just about creating information on its own but also about how businesses may get knowledge from systematic experimentation and trial and error (Figure 2.12).

FIGURE 2.12   The use of the scientific method to explore what we believe to be factual.

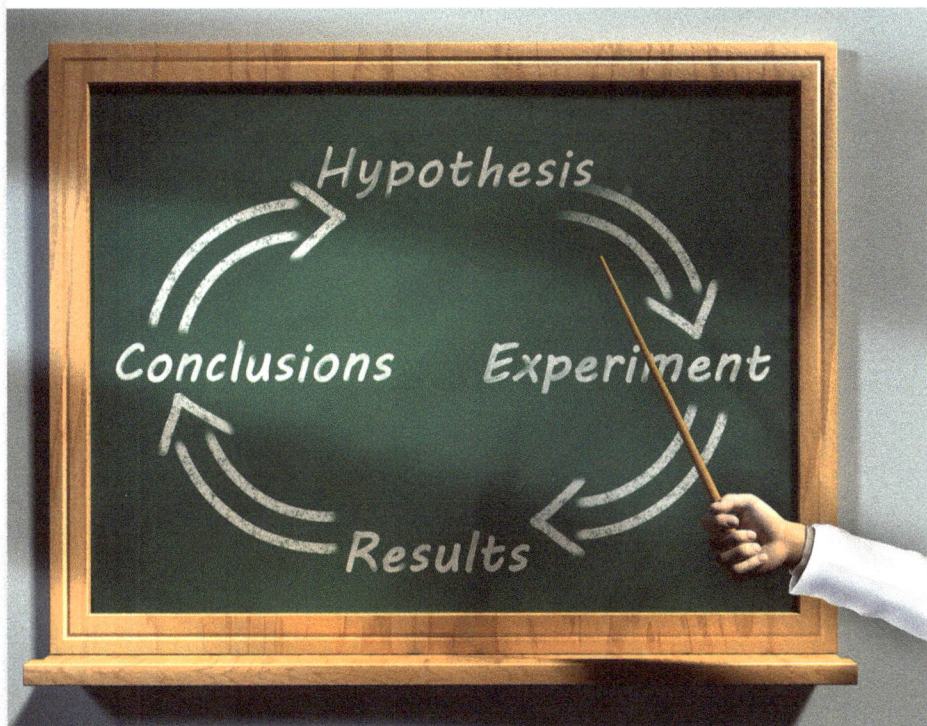

Andrea Danti/Shutterstock.com.

Design of Experiments

Design is the process of developing a strategy for the manufacture of an artifact with the purpose of resolving a problem. It is almost probable that the first act of design was user design in the sense that the plan was developed by the user rather than by a designer working for a third party.

While designing simulations and models, it is important to pay attention to the customer's instructions to create a design that is in accordance to what satisfies the customer's needs.

References

1. Masadeh, M., "Training, Education, Development and Learning: What Is the Difference?" *European Scientific Journal* 8, no. 10 (2012): 62-98.

2. Merriam-Webster, "Learning," 2018.

3. De Houwer, J., Barnes-Holmes, D., and Moors, A., "What Is Learning? On the Nature and Merits of a Functional Definition of Learning," *Psychonomic Bulletin & Review* 20 (2013): 631-642, doi:10.3758/s13423-013-0386-3.

4. Kleiner, A. and Senge, P.M., *The Fifth Discipline Fieldbook* (London, UK: Nicholas Brearley), 1994), 10, 236.

5. Schön, D., *The Reflective Practitioner: How Professionals Think in Action* (New York: Basic Books, 1983).

6. Boud, D., Cohen, R., and Walker, D. (Eds), *Using Experience for Learning* (Bristol, PA: Open University Press, 1993).

7. Draper, N., "Fail Fast: The Value of Studying Unsuccessful Technology Companies," *Media Industries* 4, no. 1 (2017): 1-19.

8. Forbes, "The Foolishness of Fail Fast, Fail Often," last accessed May 25, 2022, https://www.forbes.com/sites/danpontefract/2018/09/15/the-foolishness-of-fail-fast-failoften/#5cdf8744.

9. Khanna, R., Guler, I., and Nerkar, A., "Fail Often, Fail Big, and Fail Fast? Learning from Small Failures and R&D Performance in the Pharmaceutical Industry," *Academy of Management Journal* 59, no. 2 (2015): 436-459, doi:10.5465/amj.2013.1109.

10. Florén, H., Frishammar, J., Parida, V., and Wincent, J., "Critical Success Factors in Early New Product Development: A Review and a Conceptual Model," *International Entrepreneurship and Management Journal* 14, no. 2 (2018): 411-427.

11. Highsmith, J., *Adaptive Software Development: A Collaborative Approach to Managing Complex Systems* (San Francisco, CA: Addison-Wesley, 2013).

12. Thimbleby, H., "Delaying Commitment (Programming Strategy)," *IEEE Software* 5, no. 3 (1988): 78-86.

13. Sterman, J., "Modeling Managerial Behavior: Misperceptions of Feedback in a Dynamic Decision-Making Experiment," *Management Science* 35 (1989): 321-339.

14. Leonard-Barton, D., *Wellsprings of Knowledge* (Boston, MA: Harvard Business School Press, 1995).

15. Garvin, D., *Learning in Action* (Boston, MA: Harvard Business School Press, 2000).

16. Reinertsen, D., *Managing the Design Factory* (New York: The Free Press, 1997).

17. Fisher, R., "Studies in Crop Variation: I. An Examination of the Yield of Dressed Grain from Broadbalk," *Journal of Agricultural Science* 11 (1921): 107-135.

18. Fisher, R., "Studies in Crop Variation: II. The Manurial Response of Different Potato Varieties," *Journal of Agricultural Science* 13 (1923): 311-320.

19. Box, G. and Draper, N., *Evolutionary Operations: A Statistical Method for Process Improvement* (New York: Wiley, 1969).

20. Box, G. and Draper, N., *Empirical Model-Building and Response Surfaces* (New York: Wiley, 1987).

21. Clausing, D., *Total Quality Development: A Step-by-Step Guide to World Class Concurrent Engineering* (New York: ASME Press, 1993).

3

More Than Engineering

Creativity

One of us has ceded many patents to the companies we have worked for. The starting point for each was some technical or customer difficulty. Exploring these limitations and how to remediate them are good starting points from experience. I am not a smoker, but I found, for at least four of the patents, significant time spent outside in the smoking area with the other team members ruminating about obstacles and some ideas we could explore to surmount. Time away from the laboratory and the immediate work effort, from experience, can help you think of things differently.

Creativity and PD have become fundamental requirements in organizational environments. As they go hand in hand, they have continuously demonstrated their significance in achieving a competitive advantage in dynamic marketplaces. The application of emerging technology requires time and understanding of the available technology. The manner of application of technology can be challenging when that technology is little known or understood.

PD has become a key means of survival for many organizations and a strategy for continued profitable growth of the organization. New safety features, vehicle electrification, and alternative applications represent growth opportunities. At the core of PD is creativity. Which then brings us to the next question: What is creativity? (Figure 3.1)

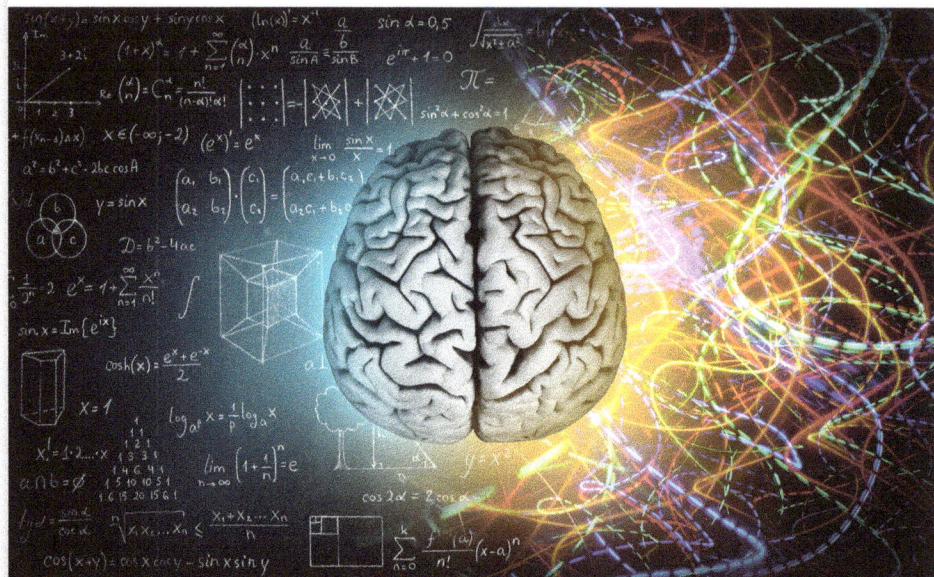

FIGURE 3.1 Engineering skill is not enough; we will need creativity to solve our difficulties.

According to one definition, creativity is "the propensity to develop or identify ideas, alternatives, or possibilities that may be valuable for problem-solving, interacting with others, and amusing ourselves and others" [1].

Webster's dictionary describes creativity as the ability to design and bring something new into existence through the power of imagination. We would add for team technical prowess when it comes to developing new products for vehicles.

In this chapter, we will explore creativity and how it correlates with the PD process, including ways of harnessing creativity and the creative tools needed for the successful completion of PD.

Importance

There is no doubt that creativity is the most important human resource of all. Without creativity, there would be no progress, and we would be forever repeating the same patterns.—Edward de Bono

Innovation and creativity are essential for solving problems. It does not matter if it is the customer's problem, finding a design solution that meets the customer's problem or a regulatory requirement, or on how to produce the product effectively and efficiently off the manufacturing line.

Companies that foster a culture of creativity encourage their employees to devote time and energy to developing novel ideas, concepts, and methods. As we will see in future chapters, some of this creativity is used to explore emerging technology and its application to the vehicle.

Today's businesses work in a highly competitive global market, making innovation an essential component of success. Creativity is the engine that discovers and drives excellent ideas, which challenges employees' thinking and opens the door to new business opportunities. Creative ideas are fodder for exploration and experimentation.

Critical Thinking an Essential Component of Creativity

Abilities in critical thinking and creative thinking are the basic skills involved in the process of forming judgments and problem-solving. Creativity and critical thinking work side by side to provide solutions and develop PD innovations.

When it comes to finding solutions to any issues, critical thinking helps analyze material and uncover the root of the underlying nature and features of problems; nevertheless, creative thinking moves the needle ahead in making progress. People who are exceptional at creative thinking can conceive or conjure new solutions—through critical thinking—to existing issues that do not rely on previous solutions. Creativity alone might be for artists; creativity in solving challenges that is engineering.

When everyone else is still debating between solution A and solution B, there are those who come up with option C. Creative thinking skills involve using strategies to clear the mind. We generate ideas beyond the current limitations of a problem and that allows us to see beyond the often-self-imposed barriers that prevent new solutions from being found. We are not fixated on a single answer or approach. Not being fixated on one solution enables us to consider various solutions to problems we have not thought of before.

The capacity to think critically about a subject, that is, to dissect a question, scenario, or issue into its component elements, enables us to evaluate the veracity of assertions, claims, and information read and heard. When distinguishing between truth and fiction, sincerity and deceit, and accurate and misleading information, the sharp knife (critical thinking) is the tool that, when sharpened, does the job. Critical thinking is a capability that is practically utilized daily by all of us, albeit to varying degrees.

In the creative process of PD, there are four stages of creative problem-solving, which all involve critical thinking. Divergent and convergent thinking are essential parts of thinking that make this process possible.

Divergent thinking is developing many thoughts in response to a single prompt. However, readers may already be aware of some tools and methods, like brainstorming, that strive toward diverse thinking.

Convergent thinking is selecting a single concept from a pool containing many others based on a set of predetermined selection criteria. Convergent thinking is illustrated by the evaluation strategies used to administer a multiple-choice standardized test.

Types of Creativity

A neuropsychologist named Arne Dietrich, who studied the neuroscience of creativity, identified four types of invention. According to his research published in 2004 [2], the four types of creativity are as follows:

1. Spontaneous and Cognitive Creativity
   When the conscious mind stops functioning and goes to relax, the unconscious mind is allowed to do its job, which is when spontaneous and cognitive creativity may occur.

This form of creativity arises when an individual possesses the skills necessary to complete a particular task but lacks the inspiration and guidance required to go in the proper direction.

In most cases, spontaneous and cognitive creativity is associated with "Eureka!" moments and happens during the most unexpected times. A good example would be how Isaac Newton discovered the law of gravity when an apple hit him on the head. Though there is a considerable possibility, there were significant experiences and learning prior. The region of the brain responsible for unconscious activities, known as the basal ganglia, takes over the efforts of the conscious brain to find a solution to the problem.

Indulging in a variety of activities that are not linked to one another allows the unconscious mind to connect knowledge in novel ways, which can lead to the discovery of solutions to issues. Therefore, to facilitate this kind of creative thinking, one needs to step away from the subject and take some time off to give the conscious mind a chance to take control. For example, as a development engineer, I would sometimes get stumped by how to achieve a software objective. How can we technically achieve this objective? Sufficiently stumped, I would pack my tent into the car and head for the mountains with the windows down and Jimmy Buffett up. But, unfortunately, I would have a couple of possible solutions to explore before I reached my destination.

2. Deliberate and Cognitive Creativity

This type of creativity comprises the intentional forms of creative people, deliberate and intellectually based. They have a wealth of information regarding a specific topic and can use that knowledge, along with their talents and capabilities, to develop a strategy.

Research, experimenting, and problem-solving are typically areas of expertise for those possessing this creativity. The prefrontal cortex, located in the front of the brain, is responsible for this sort of creative thinking.

This quadrant is typically symbolized by the well-known inventor Thomas Edison, to whom we owe a great deal of gratitude for the advancements that have been made in the fields of electricity and communications. Edison was famous for the amount of time he devoted to evaluating the results of his experiments and his dogged determination to achieve his objective.

3. Spontaneous and Emotional Creativity

The amygdala region of the human brain is responsible for the person's capacity for spontaneous and emotional creativity. The amygdala is the part of the human brain in charge of all emotional responses.

When the conscious and prefrontal regions of the brain are relaxed, the creative process and spontaneous thoughts might emerge. This form of creativity is most commonly seen in great artists, such as painters, musicians, and authors, among other creative fields. There is a connection between "epiphanies" and this form of creative thinking.

The word "epiphany" refers to a sudden insight into something. Scientific discovery, religious insights, and philosophical insights can all be attributed to the spontaneous and passionate creativity of the discoverer. This insight enables the individual who has attained enlightenment to look at an issue or circumstance from a different and more profound perspective.

Those are some of the most precious times since they are the ones in which significant discoveries are made. It is unnecessary to have any particular knowledge for "spontaneous and emotional" creativity to occur. Still, there should be a talent present, such as being able to write, play an instrument, or create art.

4. Deliberate and Emotional Creativity

People who fall into the category of being deliberate and emotional creatives allow their emotional state to impact the work that they do. Therefore, these creative persons tend to have high levels of emotion and sensitivity in their personalities. People like this like to think and reflect at times that are more private and relatively quiet, and they frequently keep journals or diaries. But, on the other hand, when it comes to making decisions, they are equally logical and sensible.

They always ensure that their creative output is harmonious with conscious emotional reflection and rational problem-solving. Parts of the human brain, the amygdala and the cingulate cortex, are responsible for this kind of creative thinking. The amygdala is the part of the brain that controls human emotions, whereas the cingulate cortex is involved in learning and processing information. People have a creative burst like this at very unexpected times.

When someone suddenly thinks of a solution to some problem or thinks of some inventive idea, those moments are typically referred to as "aha!" moments. However, for deliberate and emotional creativity to occur, one must seek some quiet time.

Traits of Creative People

Individual creativity in a person is linked to the person's styles of thinking, knowledge, and intelligence, their form of motivation, and, lastly, the person's environmental conditions [3]. In Sternberg's theory, he expressed the creativity of a general person as follows:

$$C = f(I, K, TS, P, M, E)$$

where
   C stands for creativity
   I stands for intellectual abilities
   K stands for knowledge
   TS stands for thinking style
   P stands for personality
   M stands for motivation
   E stands for Environment

Individuals tend to portray creative traits in a particular field through their personalities. This means that if a person is naturally good at something, one will be able to see that personality trait in their willingness to take risks. However, these personality traits are hard to notice in PD because the work is performed as a team activity. This is beneficial for an organization, but as a result, it downplays some individual talents.

Motivation plays a part in creativity in that it helps fulfill an individual's wish to innovate. In PD, motivation is mainly nurtured via positive reinforcement. When it comes to creativity, the environment plays a very important role. Not just the immediate environment where innovation is needed but also an individual's environmental background. An individual's immediate environment can invoke creativity or squelch it. However, the environmental background, where a person comes from, and how they grew up, also influence an individual's ability to create.

For instance, if a child grows up with a father who is a mechanic, they may pick up those skills in their childhood. Therefore, in that child's adulthood, mechanical creativity may come more easily to them than to those learning it as adults.

Knowledge also plays a significant role when it comes to creativity. Experiences, good and bad, bring about knowledge, and what better way to learn than from experiences, right? However, according to Von, O.R. [4], knowledge can sometimes be limiting. He stresses from the quote: "Imagination produces creation" that knowledge alone is not enough.

For an individual to fully maximize their creativity, knowledge is combined with imagination. A person who knows best at problem-solving because of their experiences can hardly see the solution from a different perspective.

> Imagination is more important than knowledge. For knowledge is limited to all we now know and understand, while imagination embraces the entire world, and all there ever will be to know and understand.—Albert Einstein

According to Bono, E. [5], creativity seems to go together with intelligence for individuals with an IQ of up to 120. However, anything above that, there is a divergence between intelligence and creativity. People who are known to be creative are often thought-provoking, curious, and have a variety of odd thoughts and perhaps behaviors. There are occasions when the folks involved are not even aware of what they are doing or how significant the innovation is. As a result, they frequently invent novel concepts, the likes of which cause people to be astounded.

People who have made significant self-discoveries see the world with a new perspective and have fresh and insightful thoughts. These individuals are responsible for making original discoveries which are kept secret from the rest of the world—people who have made significant contributions to their fields and have gained global recognition for their work. People who fall into this category include artists and inventors.

Creative people have a high level of physical energy in addition to a high level of cerebral energy. They put their vitality to use by coming up with original concepts. These are the types of people that put in a significant amount of time alone to reflect and contemplate.

According to Csikszentmihalyi, M. [6], creative people portray a high level of discipline. However, they are also known to be playful. The two characteristics mixed up can be described as a combination of responsible people and irresponsible simultaneously.

In addition, they occupy the extremes of the spectrum of personality traits known as extroversion and introversion. To put it in layperson's terms, creative people are highly adaptable and can have contradicting traits depending on the environment to which they are exposed.

The Myriad of Small Things

Small actions (myriad of small things) add up over time. Learning and creativity go hand in hand—knowing more gives us more material to consider, more ideas, and perhaps a more significant association between ideas and concepts. To select the best idea, we should put a plethora of ideas from which we explore. We will discuss this further in set-based development.

Small chunks of knowledge help to understand the problems at hand and relate them to past experiences to solve them successfully—small chunks of effort, small chunks of learning, in the service of an atypical idea.

Platform Thinking

We can say that platforms are a contrast of innovation. Platform thinking is a revolutionary new method of thinking about how markets function, specifically how customers, businesses or brands, other businesses, and other market participants interact and produce value [7]. A platform is a baseline product from which adaptations to a variety of customers are readily possible.

The traditional approach to understanding markets focuses on producers' and customers' roles. Value is created by optimizing various business processes, including sourcing, design, production, branding, marketing, sales, and service.

The distinct ways rivals go about their business give each of them a distinct edge. The customer is willing to pay the price for the value created by the producer of the good or service. Platforms can shorten delivery time and reduce material costs via material volume procurement and optimized manufacturing processes. This kind of thinking is also known as the pipeline model or pipeline thinking. It gets its name because a corporation competes by controlling and creating value at each stage of the activity, pipeline, or value chain.

Creativity in a Smaller Organization

Most research on organizational creativity have been conducted on larger businesses. This is because a business must have sufficient assets to explore these alternative solutions, delaying revenue stream and accounting for the time to explore alternative solutions and innovation. On the other hand, several studies have demonstrated that employees report higher levels of creative activity while working for younger, less-established companies [8]. The degree of difficulty associated with being creative often increases in proportion to the organization's size. Unfortunately, this does not appear to make any sense at all.

If more individuals are working for the firm, there is a greater chance that they will have brilliant ideas that might turn into groundbreaking technologies. Therefore, why is there such a contradiction?

First, innovation in smaller companies is an essential thing to guarantee growth. As a result, smaller companies put more effort into innovation than larger organizations. On the other hand, when a business expands, the amount of work it must do daily rises. As a result, they need to close more sales to make enough money to cover their fixed expenses.

Reorganizing or modifying activities that initially contributed to success is likely viewed as more secure than what new and inventive activities might deliver because of the importance placed on money in the short run.

Additionally, a smaller organization has fewer employees than a larger business. For example, according to the European Commission, 2015, a midsized company has less than 250 employees, while a smaller company has less than 50 employees. As a result, employee interaction is excellent and more frequent, and therefore the flow of ideas is frequent, as well as employee togetherness, which promotes productivity at work. In larger firms, on the other hand, employees rarely get a chance to interact.

How to Harness Creativity

The beginning of innovation is marked by creativity. However, creativity is not a switch one can turn on and off. As we have noted earlier, innovation does not occur in a void; rather, it is the result of organizational culture and necessitates the use of practical frameworks and methods, with incentives playing a crucial part.

Successful organizations create an environment that fosters open ideas to create room for creativity to flourish in the workplace. These organizations generate cultures that inspire staff and preserve inventive workplaces; those that fail are huge organizations that strangle creativity with rules and allow no wiggle room for adjustment. Management does play a direct part in the creative process; however, their job is not to set the environment and manage the process.

Below are some of the methods and techniques used to evoke and harness creativity:

Brainstorming

The word "brainstorming" was first used by Alex Osborn, the creator of the Creative Education Foundation, in the 1940s. At the same time, he established the concept of creative problem-solving, which is how the word brainstorming was born. We have often used brainstorming for development process improvement and material cost reductions to great success.

Alex Osborn and Sid Parnes created what is now known as the Osborn-Parnes Creative Problem-Solving Process. This paradigm is still a valuable method for the resolution of problems, even though it is rather old. The creative problem-solving process is useful in that it helps separate convergent from divergent thinking in the PD process.

When coming up with innovative concepts or methods, we frequently employ a mix of the two approaches of thinking. However, utilizing both simultaneously might lead to imbalanced or prejudiced judgments, which can impede the production of new ideas.

Brainstorming can be used to harness creativity because it is an activity that allows people to openly share their thoughts and ideas without any fear of judgment or any competition involved. Even if the thoughts or ideas might appear absurd, brainstorming opens the door to perhaps radical alternatives that can be analyzed to determine how to use them if possible. This is a place for innovation to begin.

Three Steps of Harnessing Creativity through Brainstorming in a Work Environment

Step 1: Prepare the Development Team In this step, the project manager should prepare the participants for the brainstorming event. Not necessarily about the topic, but the project manager needs to select the people who will be present and inform them of the time and venue. They should also prepare a room for all the present members to feel comfortable.

Step 2: Present the Problem At this stage, the project manager should clearly explain the issue the team is attempting to solve and describe any criteria that must be adhered to succeed. It should be sure that everyone understands that the purpose of the meeting is to come up with as many new ideas as possible.

At the beginning of the meeting, the project manager should give everyone plenty of alone time so that they may jot down as many of their thoughts as possible. Then, after providing everyone an equal opportunity to contribute, they should invite them to share their ideas with the group.

Step 3: Guiding the Discussion The group facilitator should maintain order throughout the discussion by keeping it to one person talking at a time. The facilitator should additionally contribute their ideas and spend time supporting other team members through their sharing sessions.

When everyone has finished sharing their thoughts, the facilitator should begin a group discussion to build upon the ideas of others and use those thoughts to generate brand-new concepts. When coming up with ideas as a group, one of the most beneficial things to do is to build on the ideas of others.

The facilitator should encourage everyone, including the most reserved individuals, to contribute and come up with ideas and discourage anybody from criticizing the ideas that others have developed.

In conclusion, people should refrain from applauding or condemning ideas during brainstorming sessions. You are attempting to eliminate erroneous presumptions about the boundaries of the issue and open new avenues of investigation. Judging and analyzing something at this point might stifle creativity and prevent the production of new ideas. At the end of the meeting, you should evaluate the ideas discussed; now is the time to investigate the problems using more traditional methods.

Negative Brainstorming

Another word for negative brainstorming is reverse brainstorming. This technique encourages the team to brainstorm the problems that may appear in the innovation rather than looking at the bright side. Consider it risk exploration in service of PD and akin to the premortem sometimes used in project management to evoke risks. The problem with a topic or proposal is investigated using this line of reasoning, and its weaknesses, hazards, and hurdles are explored. Judgment and prudence are used throughout the process [9].

The reasoning behind negative brainstorming is that if you can anticipate possible challenges before beginning a project, you will be in a better position to tackle such challenges and be successful. Negative brainstorming is surely one of the most effective strategies for moving a team from resistance to triumph despite obstacles.

In contrast to traditional brainstorming, reverse brainstorming focuses on potential causes of or methods to exacerbate problems. After that, you turn these concepts around to come up with new ideas for solutions. To apply this strategy, you must first ask yourself one of the following "reverse" questions:

Instead of asking, "How can I find a solution to this issue or avoid it from happening again?" Ask yourself, "In what ways may I potentially be contributing to the issue?" Asking yourself, "How can I possibly produce the opposite effect?" is a better question than "How can I possibly attain these outcomes?"

After that, develop potential solutions to the problem by engaging in creative thinking. Let the thoughts flow freely, and avoid dismissing anything that is suggested at this point.

After you have come up with all the potential causes of the problem through brainstorming, it is time to turn these causes into possible solutions to the initial dilemma.

Finally, we will assess these potential solutions. Do you think there could be a way out of this? Do you have any ideas about what would make a good solution?

You can reverse brainstorm on your own, just like regular brainstorming, but it is more likely that you will come up with more diverse ideas if you do it in a group. Trying reverse brainstorming is a good idea when it is challenging to come up with straight answers to the issue. It can reveal subtle flaws in a procedure or product.

The Insights Game

Have you ever had one of those "light bulb" moments where you suddenly understand how something works or how the different bits in the puzzle fit together? We seek to find our eureka moment similar to Archimedes and the proverbial bathtub! These moments are the goal of The Insights Game—what the insight game aims to achieve.

As we age, our brains get better at putting together different pieces of information, seeing new patterns, and putting old ideas to the test with details. Unfortunately, as adults, we cannot memorize and learn new things as quickly as we could when we were kids. However, because we have more life experiences, we can use a broader range of information to see the big picture. The insights game is the method that can be used in the PD process (and product) to help the development team get better at coming up with new ideas and seeing new patterns.

The basic idea is to take a moment to think about new information and see if it fits into any patterns you already know or if it makes you think differently about something you already thought you knew.

When you have new thoughts, your brain makes new connections. When your brain has more connections, it can do more complex problems. This is a good reason for a development team to keep playing the insights game to harness their creativity.

This technique/game is intended for a single individual. However, it can also be used with many people or even a whole group. You may use reflection questions to assist individuals in starting to practice seeing the broad picture and challenging their previous viewpoints.

How the Insights Game Is Played   You earn one point for every insight that you provide. If you do not get at least one new perspective each day, the game ends, and you have to start over with 0 points. If you do, you will be able to continue playing. In addition, you are awarded bonus points for any insights that cause you to reconsider your prior position on a topic. You will be rewarded for playing this game if you can enhance your capacity to analyze more complex situations, perceive the bigger picture, and challenge your views; this is the game's purpose. You have successfully finished the game once you have attained nirvana. Nirvana is the final stage of the game. The name is used as a metaphor to show the successful completion of a thinking process of a project.

In its origin, nirvana is a term commonly utilized in the Buddhist community. It means a transcendental stage in which there is no desire, sorrow, or feeling of self, and the subject is liberated from karma and the cycle of death and rebirth. When you get a new thought or insight, scribble it down and give yourself some time to consider what it means and how you might put it to use. Permit yourself to cheer yourself on your victories in the insights game.

On the other hand, do not ignore the enlightening moments that come to you. It makes no difference how little the discoveries are, they are all useful. The insights game is an excellent game to use in the PD process because, first, it breaks monotony at work and helps channel creativity for innovations and problem-solving.

Mood Boards

A mood board, also known as an inspiration board, is a tool used at the beginning of a creative endeavor to help one refine their visual concepts. The person or team will be guided in their work by a collage of pictures, examples of materials, color schemes, and occasionally descriptive phrases and typography.

Mood boards are created by designers, project managers, illustrators, development teams, photographers, filmmakers, and other creative professionals to convey the "feel" of a concept. Mood boards are always beneficial, especially in the development, testing, and differentiation activities throughout the proof-of-concept phase. If a verbal explanation would be too complicated to provide, then mood boards are the next best thing. During presentations, they assist the audience in obtaining a high-level understanding of a complex product concept in a short amount of time.

According to Lucero [10], the construction of a mood board is a common practice in the fashion and textile design industries. This practice assists the designer in determining a starting point at the beginning of the production of a collection. We see no reason why this approach should be limited to textiles and fashion.

In design and innovation, mood boards are a frequently employed tool [11, 12]. Rather than providing a visual collection of actual objects, a mood board is an abstract method of grounding the design process.

During the design process, one technique that can increase both the flow of inspiration and communication is the use of mood boards. They serve as a source of inspiration for the development team, a platform for discussion and communication with the customer and other stakeholders, including users.

Random Words (Random Input)

When we cannot generate any fresher ideas, we risk being frustrated, which causes us to waste both time and the ability to concentrate. However, you may find your way out of the gloom of dissatisfaction and get closer to achieving the ultimate goal you have set for yourself with the assistance of a straightforward and speedy creative method known as random input.

Random input is a method of lateral thinking coined as a creative thinking approach. It is a technique that assists in generating fresh ideas and perspectives throughout the problem-solving process. Random input comprises the act of using random words, sounds, or even images to generate new ideas. The purpose of employing the random input strategy is not to immediately find a solution to the problem; instead, the goal is to assist the mind, in the beginning, to think in a different direction and explore creativity in new areas by using random phrases.

Random input is one of the top creative thinking techniques now used in online and physical classroom structures. It was first associated with the lateral thinking programs Dr. Edward de Bono developed, but it has since become one of the most popular creative thinking techniques overall.

The use of the random input technique has been shown to be beneficial in various contexts, including the promotion of practical problem-solving, the discovery of novel approaches to improving enterprises, and the innovation and creativity of new ideas.

In its most basic form, random input functions in a manner analogous to the method of brainstorming. On the other hand, while using the random input approach, the goal is to discover as many answers to the question as possible, regardless of whether or not those answers are correct. This is the principle of the technique. The purpose of both lateral thinking and random input is to encourage people to think in ways that are different from the norm.

How to Use Random Words to Harness Creativity

- The use of random input is possible in either the dynamic of a team or an individual. However, before beginning to use the method of random input, one must first have a well-defined issue.

- A random term, phrase, or even a single image is selected once the fundamental problem has been discovered. Recognize that the method's effectiveness is dependent on the unpredictability of the word. It is now possible to generate comments and sentences at random using various Internet programs and websites, which adds credence to the effectiveness of this strategy. Using technology is one of the quickest and easiest methods to get started in a new creative endeavor.

- Observe the associations and functions of the created term and, if appropriate, use metaphorical features of it to conjure a hypothetical situation. The information gained through brainstorming around this particular term is analyzed and applied to the original problem to find a solution. Associative thinking frequently yields unexpected results.

In some cases, connecting the random input correlations to the underlying problem may be challenging. Nonetheless, the new ideas that were generated by the thought process will lead to valuable solutions at some point in the future.

Storyboarding

Storyboarding is a technique that depicts the dynamic character of events or ideas through sequences of drawings or diagrams (Figure 3.2). Initially, storyboarding has been utilized most frequently in the film business as well as the advertising sector.

FIGURE 3.2   To envision how the product is used, we interview customers and storyboard the application.

The practice of storyboarding, as we know it, began in the late 1920s in the animation studio of Walt Disney. As time went on, the method became commonplace in various additional films and animation businesses. Annette Michelson, a writer and art curator, views the period of the 1940s as the period in which "the adoption of the storyboard has predominantly characterized production design."

In the practice known as "storyboarding," different scenes are generated from a narrative or event to facilitate the study and understanding of the material. This is something that is done frequently in a variety of academic and professional contexts. It is akin to the technical walk-through of a proposed product or system using graphical techniques.

Storyboarding is a technique that is frequently utilized in the early phases of film production, for example, by filmmakers. The many scenes are summarized so that most of the pieces that make up each scene may be read about, viewed, and debated [13, 14]. In other words, storyboarding helps to bring a story to life. In addition, storyboards, as opposed to traditional word representations of scenarios, enable visual mental activity, stimulating creative processes.

The use of storyboarding nowadays helps to enhance group brainstorming and communication among employees. A group will generate more ideas, and frequently more creative ones, if they have access to a substantial physical canvas on which graphically arrayed information is displayed. The graphical character of storyboards allows for a detailed step-by-step exposition of the thought process that goes into a procedure from the beginning to the end.

Metaphorical Thinking

There is a strong correlation between symbolic thinking and increased creativity; alternatively, it may be more accurate to state that we can employ metaphorical thinking to locate original ideas and insights. A metaphor is a comparison between two objects that are not directly related to one another but are connected in some other way. In ordinary communication and thought processes, metaphors may be extensively employed. It is, therefore, possible to use metaphors to enhance communication in PD.

Linking two concepts together by using a metaphor enables the merging of components with very little to no logical connection to one another. In this way, metaphors can open up the creative side of the brain, which is partly stimulated by pictures, ideas, and concepts.

Metaphors defy the laws of logic in this way. Therefore, thinking of problems in metaphorical terms might help for creative issue-solving: It encourages one to "think outside of the box" to use another well-known term. The use of metaphors allows individuals to view the world in a whole different reflection and come up with creative solutions: possibly solutions that never even existed.

Mind Mapping

First and foremost, a mind map is categorized as a visual thinking tool. Unlike mood boards, mind-mapping activity is associated with a person's free flow of thoughts and ideas and associations to those thoughts.

When creating a mind map, one starts by putting down a core thought and then works their way outward by thinking of other concepts connected to the original idea. To better grasp and remember information, try mapping it by focusing on the main points put down in your own words and looking for links between them.

Mind mapping is a beneficial way of harnessing creativity in innovation. Using visual stimuli in conjunction with your thoughts and ideas can, as a result, assist your brain in processing information more quickly and drawing more creative correlations between your ideas.

Lateral Thinking

Edward de Bono originally used the phrase "lateral thinking." The perception is at the center of the thought process for lateral thinkers. The outside world is broken down into manageable chunks and organized here through the lateral thinking process.

It involves addressing issues indirectly and creatively, employing reasoning that is not immediately clear and involves concepts that may not be accessible by typical step-by-step logic.

Ways to Harness Creativity through Lateral Thinking One way to get your creative juices flowing using lateral thinking is by trying on a new identity. Trying to think in a different way might be challenging at times. One strategy that has shown to be successful is to put on a separate set of mental hats and conduct your cognitive processes as though you were someone else.

The second way is to be inquisitive. If you question everything sufficiently, you could find a different angle on the subject or problem. Before attempting to find answers, it might help to ask yourself some questions about the big picture.

On a third note, a different approach would be to start with potential answers in mind and work backward. Since there is only one rule for lateral thinking, which is to "think beyond the box," the thought process can take place in any sequence. Thinking backward can effectively enable one to tap into their inner resource of imagination and think creatively outside the box.

Creative Tools

Creative tools and techniques play a very important role in the PD process. They assist the multidisciplinary teams in problem-solving and through any arising issues in the PD process, therefore helping them come up with superior products. Below is a review of various creative tools and techniques that can be used in PD.

Mind Mapping

Mind mapping was developed as a thinking tool for generating concepts and ideas through association. It is a visual method that may be used to manage complicated commitments, organize tasks, and provide documentation that is simple to comprehend.

Using a mind map allows one to construct links between several concepts or bits of information visually. The first step in this method is to write down each thought and then follow it by connecting each idea to other connected components using lines or curves. As a consequence of this, a web of relationships is created.

The consultant can determine the most effective strategy and anticipated outcomes by altering these relationships and assessing them. Development teams can record, categorize, examine, and depict complex thoughts and concepts when they use mind maps as creative tools.

In other words, mind mapping transforms mundane information into a visually appealing, easily recallable, and meticulously arranged diagram consistent with how the brain normally operates. By using mind mapping in PD, the whole team can break down complex information into understandable information, provoking creativity and a stream of ideas.

Da Vinci

It is generally agreed that Leonardo da Vinci was the greatest genius person who ever lived. He was a man whose curiosity could not be satiated and whose mind worked frenetically (Figure 3.3).

FIGURE 3.3 Da Vinci can be argued to be the pinnacle of genius.

InnaBor/Shutterstock.com.

Although the mention of his name may evoke visions of well-known works of art like the Mona Lisa, The Vitruvian Man, or The Last Supper, he was much more than just an artist. He was also an architect, a mathematician, and an engineer. The Italian maestro was not only talented and creative but also possessed another quality: the capacity to take a unique perspective on the world surrounding him. Here are a few concepts that da Vinci has taught the world about creativity:

Maintaining Curiosity and a Passion for Continuous Learning One of the most exciting concepts that da Vinci emphasized was passionate curiosity and an unquenchable thirst for learning. Therefore, it is essential to have an insatiable curiosity fueled by intellectual curiosity and a burning desire to acquire further knowledge.

This highlights the importance of being open to one's intuition and discernment and having a mind that is actively seeking knowledge. For curiosity to be a source of creative energy, it must be unquenchable, possess an open imagination, and be motivated by a wide range of different things. Passionate curiosity helps one understand other points of view and spot problem-solving techniques.

Getting in Touch with One's Senses Unleashing curiosity comes as a package deal, attentive observation skills and enhanced listening abilities. To learn more, we must be able to use our ears and eyes and listen more than we talk. According to Leonardo da Vinci, observation is the basis for creativity since it lays the groundwork for a more comprehensive framework, expands one's ability to think, and broadens their perspective.

Think carefully, reflect on it, and evaluate the significance of what you have seen. Then, trust your observations to transform you completely. Your senses will get more refined and heightened, resulting in higher comprehension and a broader and deeper perspective on the situations you find yourself.

Because of the way he saw the world, Leonardo da Vinci was able to produce some of his greatest works of art and find breakthroughs in his scientific discoveries.

Balancing Logic and Imagination in the Thought Process The thought processes of Leonardo da Vinci were superior to those of his contemporaries in many ways. When deciding, his approach was more organized, neutral, and creative.

This allowed him to keep an eye on two conflicting platforms at once, spot commonalities among them, and then combine the resulting facts and statistics in a way that trusted both his logic and his intuition.

You are putting together logic and creativity, resulting in a process that is not only thrilling but also constructive. You will be able to organize your thoughts better and come up with more ideas in a shorter time if you use the mind-mapping technique.

Thinking Big and Dreaming Big   When it comes to thinking big, it is essential to think beyond and outside the box. To do so, one must give up what is comfortable and predictable. It is about looking behind rather than forward (working from the future toward the present and not the other way) to make better decisions.

Creative intellect is unleashed, and the greatest number of possibilities are found, with no ceilings or biases and no preference for the better ones. Leonardo da Vinci's teachings emphasize thinking broadly before diving into a project's intricacies.

Making New Connections and Looking for Patterns   Taking a step back and looking at the big picture first helps you see patterns and discover new connections. When it comes to Leonardo da Vinci, everything is connected. He was confident that everything had a purpose. According to him, the best way to approach a problem is to seek for patterns, reasons, parallels, intersections, and consequences.

Discard old ideas by generating new ones in your head. To generate new ideas, try reverse brainstorming. Mix and match a concept or product with something completely random. Then, list the similarities and differences between the two. Dissimilar items can be compared and merged to create a new product.

Seeking Out Different Perspectives on Matters   Your creative potential will be significantly stunted if you limit yourself to considering only a single perspective. Engaging in conversation with other people and taking into consideration their views, in addition to your own, is an essential part of intelligent work.

Embracing Uncertainty with Confidence

> The creative person is flexible; he is able to change as the situation changes, to break habits, to face indecision and changes in conditions without undue stress. He is not threatened by the unexpected as rigid, inflexible people are.—Frank Goble

The mystery of Leonardo da Vinci's work can be attributed to his willingness to embrace ambiguity, contradiction, and uncertainty.

It is essential to engage in what is known as the Socratic Method, which involves questioning without preparing definitive responses in advance. The ability to have self-assurance even when the relationships between causes and consequences are unclear is an essential component of creativity.

Six Thinking Hats (de Bono and Lateral Thinking)

Dr. Edward de Bono invented the six thinking hats in the early 1980s to facilitate the possibility that problems can be discussed from various points of view. The notion of the "six thinking hats" offers individuals a framework for extending and organizing their thinking, as well as a platform for more successful decision-making (Figure 3.4).

Whale Design/Shutterstock.com.

FIGURE 3.4   We need to think creatively and Six Thinking Hats is one approach.

The key to the successful use of this strategy is to break the thought process down into six distinct areas and tackle each one at a time. This approach makes it easier to think in a concentrated manner.

Distinct ways of thinking and behaviors are represented by each of the six different colored hats. This strategy makes use of the significant functions that colors play in human existence [15].

They are embodied symbols that, when encountered, cause certain roles to be played, so allowing the teams' thinking to deviate from its typical patterns. The act of wearing a hat is intentional as each hat stimulates a certain mode of thought [16].

The Six Thinking Hats and How They Work

1. **The White hat** symbolizes a singular concentration on the current information as well as the subject at hand. It performs the role of information collector by performing research and incorporating quantitative analysis into the conversation; adheres to the facts.

   Additionally, the white hat encourages the thinker to differentiate between what is factual and what is their own personal interpretation of the facts. The more knowledge individuals acquire the better their reasoning and the more suitable their choices will be [15].

2. **The Red hat** symbolizes the ability to freely communicate feelings, intuition, hunches, and emotions while avoiding the need to provide an explanation. Logic is not of importance or required at all with the red hat thinking process.

   The red hat provides us with a different setting in which we can disclose our honest sentiments and then investigate the ramifications of those feelings.

3. **The Black hat** represents caution and prevents us from acting in ways that might be harmful, counterproductive, or just impossible. Unlike the yellow hat, the black hat provides constructive criticism and finds flaws in the idea or concept.

   For risk assessment, the black hat is most suitable as it facilitates critical thinking.

4. **The Yellow hat** is the logically positive hat, under which the thinker finds out the idea's merits and advantages and considers how it may be implemented. Everyone, in turn, is required to mention what they like about the proposition or the concept being presented. Even if one believes the proposal will not be successful, they are still obligated to look for some redeeming traits and excellent elements in it [17]. By wearing the yellow hat, the team's thinking is automatically redirected toward how the idea is doable and how they can get the best benefits out of it.

5. **The Green hat** represents our creative and action hat through which we propose alternatives, search for new ideas, and produce possibilities. It encourages the process of creatively providing solutions to the problems arising from the innovative idea.

   This way of thinking, according to Macdonald [18], offers provocations, fresh insights, and spectacular alternatives without making any attempt to critique or assess the value of these concepts. Green hat thinking may be used to upend things and embark on a new course.

6. **The Blue hat**, also known as the control hat, is one that a person wears when they are preoccupied with regulating the thought process. Putting on the blue hat is similar to having a vantage point from which one may look down on a circumstance and formulate a strategy for how one ought to think about it. It governs the thought process by choosing what the next step will be and creating the agenda for the session.

   The blue hat may be used for process control at both the beginning (for planning) and the end (for summarizing) of each session [19].

A team can concentrate their thoughts and overcome barriers to efficient practical thinking with the help of the six various colored hats that are provided. Incorporating different ways of thinking and problem-solving into the PD process is made easier by the hats.

Lateral Thinking Lateral thinking, also called horizontal thinking, is a part of Dr. Edward de Bono's work in creativity. Lateral thinking can be used as a tool for creativity in PD. In lateral thinking, the development team tackles challenges by employing reasoning that is unconventional or that is not immediately clear. In other words, it involves thinking out of the box.

> You cannot dig a hole in a different place by digging the same hole deeper.—Dr. Edward de Bono

Lateral thinking can be used as a creative tool in enabling a team to get a fresh perspective of ideas and solutions. This is most effective when applied during the early stages of divergent thinking, which is the ideation stage in PD.

The main approach to getting a fresh perspective is through provocations. Provocations are a common strategy in which intentionally incorrect claims regarding a problem or a situation are made. This may be done to call into question established standards by employing techniques such as contradiction, distortion, wishful thinking, reversal (of assumptions), or escapism.

Another way lateral thinking can be used as a tool for creativity is by helping the team to break out of thinking inside the box. "The box" is a metaphor for the seeming limitations of the design space, as well as the constrained viewpoint we have as a result of our tendency to approach problems in a frontal and sequential manner.

By engaging in lateral thinking, the development team can learn to take a step back and utilize their power of imagination. This is done to see the bigger picture rather than letting logic and preconceptions keep their thought process confined and limited.

This can be done in the following ways:

- Trying to look for different options, not simply different potential answers but also different ways of thinking about the issues.

- Paying attention to the underappreciated facets of a situation or the problem at hand.

- To break out from the conventional methods of interpreting an issue, concept, or solution, it is important to challenge assumptions.

By doing this, the team or an individual can open themselves up to disruptive thinking, as well as give themselves the ability to flip an established paradigm on its head.

The technique known as the "Six Thinking Hats" can be utilized in a variety of contexts (including PD) where it is necessary to engage in creative and critical thinking, brainstorming, and lateral thinking.

Brainstorming

Brainstorming is probably one of the most popular creative techniques (Figure 3.5). This is the most apparent creative technique, and an endless whiteboard is just perfect for it.

FIGURE 3.5   We can tap into the creativity of our team members through brainstorming.

REDPIXEL.PL/Shutterstock.com.

Brainstorming can be used as the first step (ideation stage) of the PD process, and it is a powerful tool.

A team may use brainstorming as a tool to discover innovative solutions to an issue by engaging in the act of brainstorming. This technique is beneficial when there is a need to break out of stale thought habits since it forces out new ways of thinking.

Most people have tried brainstorming at least once, which is the easiest way to think creatively on purpose. With the ability to quickly come up with many ideas at once, we can avoid the natural tendency of the brain to limit our ability to come up with solutions. This lets us access and combines many possible solutions/thoughts to come up with new ones.

The process of brainstorming incorporates a laid-back, casual attitude toward the solution of problems with lateral thinking. This tool inspires people to conceive of ideas and concepts which, at first glance, may appear to be completely absurd. Some of these concepts have the potential to be developed into novel and imaginative approaches to solving an issue, while others have the potential to stimulate the generation of even more ideas.

It is like sprinting through the finish line of a race only to find a new track on the other side and the option to keep going. Like critical thinking, PD requires us to think creatively, and the best way to practice and improve this skill is by brainstorming.

Engaging in brainstorming activities in the workplace is one method for coming up with fresh concepts and innovative ideas and getting the creative juices flowing.

TRIZ

TRIZ was developed by Genrich Altshuller [20], a Soviet engineer and an inventor. It is a scientific approach to creative problem-solving in the PD process (Figure 3.6).

FIGURE 3.6   TRIZ is another approach to using creativity to help solve the problem or constraints at hand.

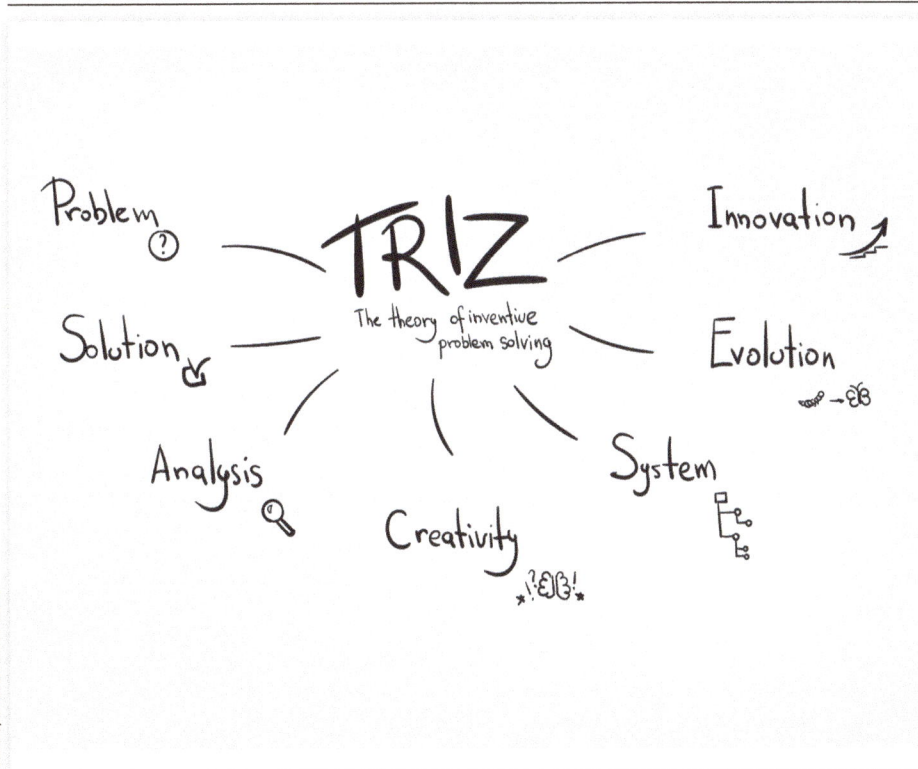

Natata/Shutterstock.com.

TRIZ originally translates to "Theoria Resheneyva Isobretatelskehuh Zadach" in the Russian language, where the approach originated from. However, it can also be translated into English as the "Theory of Inventive Problem Solving."

TRIZ is an approach to problem-solving that makes use of both divergent and convergent thinking capabilities. This approach uses the activity of coming up with creative concepts by making use of previously gained expertise in the field of human innovation [21].

A vast patent database is where this body of knowledge is accumulated and recovered from. This database is also the tool that is used to identify how the problem might be handled based on the evolution of products and technologies that are already in existence.

The primary idea behind TRIZ is to offer innovators simple access to a wide variety of experiences and knowledge of earlier inventions and leverage previous answers for tackling new arising innovative issues in the PD process [22].

TRIZ is based on the idea that every problem with a product is caused by a conflict between criteria or requirements. The problem will be solved when the conflict is taken care of.

Inspiration does not have to be random to come up with new ideas, but we can take a structured approach to tackle and control the innovative process to find new ways to solve these problems [21].

Based on his investigation of the process through which innovations develop over time, Altshuller came up with a list of 40 inventive principles and 39 features. When examining a PD issue, these are utilized to determine prospectively.

References

1. Franken, R.E., *Human Motivation* (Pacific Grove, CA: Brooks/Cole Publishing Company, 1998).

2. Dietrich, A., "The Cognitive Neuroscience of Creativity," *Psychonomic Bulletin & Review* 11, no. 6 (2004): 1011-1026.

3. Sternberg, R.J. and Lubart, T.I., *Defying the Crowd: Cultivating Creativity in a Culture of Conformity* (New York: Free Press, 1995).

4. Von, O.R., *A Whack on the Side of the Head* (New York: Warner, 1983).

5. Bono, E., *Serious Creativity: Using the Power of Lateral Thinking to Create New Ideas* (London, UK: HarperCollins, 1992).

6. Csikszentmihalyi, M., *Flow and the Psychology of Discovery and Invention* (New York: Harper Perennial, 1997), 39.

7. Choudary, S.P., "A Platform-Thinking Approach to Innovation," Wired.com, 2014.

8. Solomon, Y., "Startup to Maturity: A Case Study of Employee Creativity Antecedents in High Tech Companies," Dr. Art. thesis, Capella University, Minneapolis, MN, 2010.

9. Vialle, W., Lysaght, P., and Verenikina, I., *Psychology for Educators* (Sydney, Australia: Social Science Press, 2005).

10. Lucero, A., "Framing, Aligning, Paradoxing, Abstracting, and Directing: How Design Mood Boards Work," in *Proceedings of the Designing Interactive Systems Conference*, Newcastle upon Tyne, UK, June 2012, 438-447.

11. Cassidy, T., "The Mood Board Process Modeled and Understood as a Qualitative Design Research Tool," *Fashion Practice* 3, no. 2 (2011): 225-251.

12. McDonagh, D. and Storer, I., "Mood Boards as a Design Catalyst and Resource: Researching an Under-Researched Area," *The Design Journal* 7, no. 3 (2004): 16-31.

13. Giuseppe, C., *The Storyboard Artist* (Studio City, CA: Michael Wiese Productions, 2011).

14. Tumminello, W., *Exploring Storyboarding* (Canada: Thomson, 2005).

15. De Bono, E., *Six Thinking Hats* (New York: Back Bay Books, 1999).

16. Jensen, E. and Nickelsen, L., *Deeper Learning: 7 Powerful Strategies for In-Depth and Longer-Lasting Learning* (Thousand Oaks, CA: Corwin Press, 2008).

17. Sloane, P., *The Leader's Guide to Lateral Thinking Skills: Unlocking the Creativity and Innovation in You and Your Team* (London, UK: MPG Books Ltd, 2006).

18. Macdonald, M., *Your Brain: The Missing Manual* (Toronto, Canada: O'Reilly Media, Inc., 2008).

19. McGregor, D., *Developing Thinking: Developing Learning: A Guide to Thinking Skills* (London, UK: Library of Congress cataloging-in-Publication Data CIP dat, 2007).

20. Altshuller, G., *Creativity as an Exact Science: the Theory of the Solution of Inventive Problems* (New York: Gordon & Breach, 1979).

21. Gonzalez, D., *When We Peek behind the Curtain: Highlighting the Essence of Creativity Methodologies* (Evanston, IL: THinc Communications, 2002).

22. Moehrle, M.G., "What Is TRIZ? From Conceptual Basics to a Framework for Research," *Creativity and Innovation Management* 14, no. 1 (2005): 3-13.

4

Front Loading

Front-Load the Product Development Process

The second principle of the Lean Product Development System model, *Front-Loading the Product Development Process to Explore Alternatives Thoroughly*, is akin to the shop-floor adage of Measure Twice, Cut Once, and is the cornerstone of faultless execution throughout the program (Figure 4.1).

FIGURE 4.1   Front loading reduces risk through learning and developing the application.

Although initial plans are frequently subject to modification, the success of the lean product development (LPD) program depends entirely on the meticulous, rigorous planning process. Key to the success of the program is the advancement of an LPD system. An optimum system is one that works to strip away the waste in the planning process. An approach that brings the team together to get the best from the team of engineers.

Late changes significantly impact project costs and the profitability of the company. Figure 4.2 illustrates how front loading of engineering activities has the potential to have the most benefit for all stakeholders.

FIGURE 4.2 Cost and product change curve example.

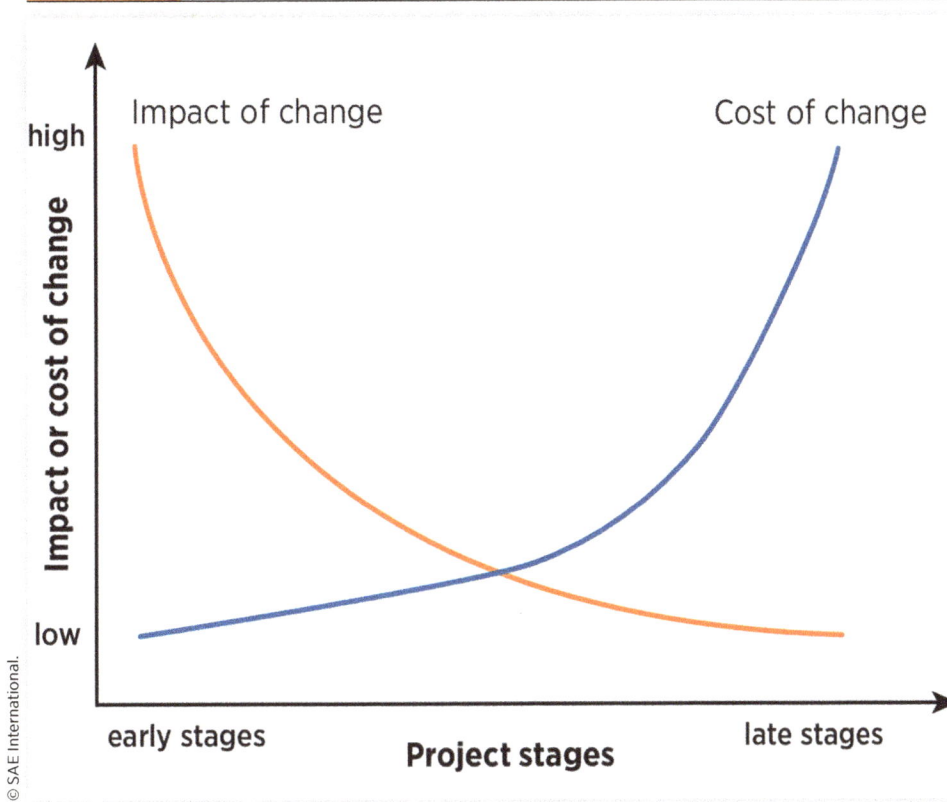

Front loading removes the classic PD problem of late design changes, which are expensive, suboptimal, and continually worsen product and process performance since they are solved at the root cause level early in the process. Late changes happen when we **prematurely** select the desired solution. Late engineering adjustments are "short fixes" or patches rather than continual progress; they are the worst type of waste. In addition, these changes cost the project dollars and time to market. Toyota attempts to isolate and decrease variation early in the process since it significantly impacts queues and other system delays during the execution phase of PD.

Front loading enables standardization of the architecture, the procedure, particular activities, and performance objectives, providing long-term benefits. Through early stage designing of the PD process to address the causes of variation, settle disagreements, solve problems, and separate them from the remainder of the PD process (Figure 4.3).

FIGURE 4.3 Front loading covers many areas.

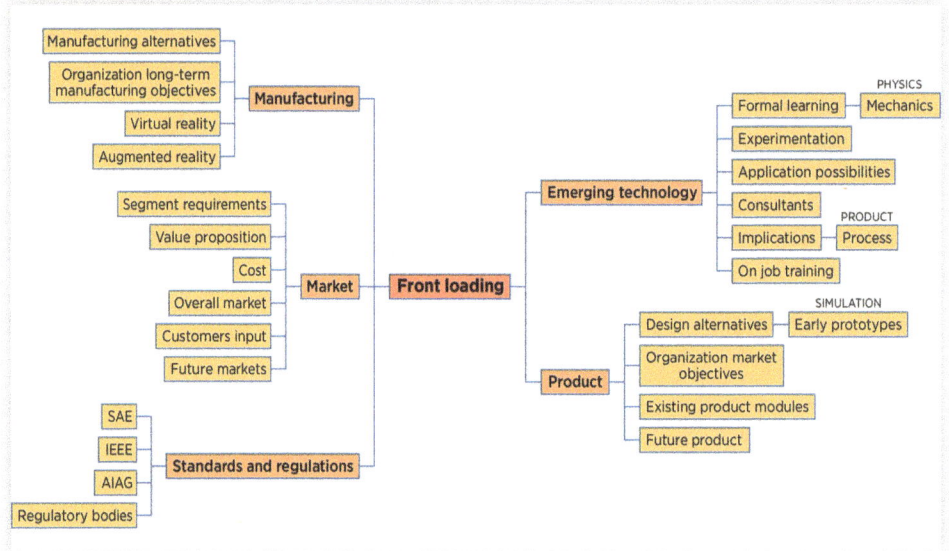

© SAE International.

Front Loading Product Development

Anyone involved in PD knows the anguish of late adjustments, sometimes called loopbacks. To "front-load" the PD process is a typical recommendation to avoid this issue. What does front-load mean though? It is complex, but we attempt to answer it below.

Front loading focuses on learning and resolving the highest priority functions and risks of the effort. We experiment and explore early. The desire is to limit rework while continuously building the product. The pain of late-stage adjustments, sometimes called loopbacks, is well known to everybody who participates in the PD process. Design or engineering changes late in the development process result in significant rework. There are many manifestations—an unexpected failure in prototype testing. Manufacturing issue that surfaces during production ramp-up or warranty issue found early in the field all result in waste and point to a lack of front loading. While development teams work extremely hard to prevent them, nearly all practitioners view loopbacks as unavoidable and even prepare for them. However, this is unfortunate from a lean perspective because a loopback could be another name for rework. The same tasks product developers previously performed: create, evaluate, and test are performed again later in the development process.

What Does Front-Load Mean?

Some have advised starting a PD project by gathering our existing knowledge. For instance, at the beginning of the project, manufacturing engineers can provide the team with details on the process capability of their present manufacturing equipment. The product designers will then be able to create within those limitations to reduce impacts on the production floor while enhancing quality and efficiency. This suggestion is difficult because learning everything there is to know is impossible. But it is also in translating that knowledge from technical terminology related to processes to ones that product designers can comprehend. Lean businesses make a substantial effort to continuously catalog their process capability so that the information is available for any projected new PD.

Others have advocated creating cross-functional teams from the beginning of development projects to front-load them. Teams are comprised of competent individuals from every department required to bring a product to market, including purchasing, engineering, product engineering, industrial design, and finance. We prevent unanticipated "discoveries" later in the process by ensuring that all points of view are accounted for from the beginning. However, one concern frequently plagues these teams: what should they do in those initial stages?

Stefan Thomke advises quick experimentation to discover the potential of products and production methods. Allen Ward's theories of set-based design, or what we today refer to asset-based innovation, allow us to expand on Thomke's concept.

Teams interact with the development difficulties they experience fundamentally different than conventional techniques while using a set-based approach to development. The following are the main methods of set-based innovation:

For each identified design subproblem, create several potential solutions and compare them to the design requirements by gathering data using analysis, modeling, and low-cost prototyping. Pay attention to the level of detail in the data required, for example, employ low-fidelity models when you need quick, rough answers.

- When the facts indicate that an alternative is either unable to satisfy the requirements or is blatantly inferior to another option, we should eliminate it. Instead of picking a winner right away, adopt an elimination procedure.

- Through design constraints and tradeoff curves, try to produce reusable knowledge along the way. You can confidently rule out an alternative if you are aware of the physical limitations of a design concept and those limitations exceed the criteria.

- Be flexible with your design needs so that you may determine the final requirements once you have a better understanding of the trade-offs involved.

System Dynamics: Investigating Front-Load Effects

Over the last few decades, fierce competition has caused new items to be viewed as a worldwide conflict. Therefore, businesses never stop trying to enhance and develop new tactics, engineering processes, and approaches. Various applications of the well-known lean philosophy have had success and failure, depending on the context for NPD and the requirements of the philosophy itself. The Toyota approach, which Morgan and Liker have extensively discussed, is one tried-and-true way to compete and succeed in the fast-paced and competitive automobile business. However, in a broader sense, competing with new products necessitates significant ongoing efforts and drastic adjustments. This may entail altering strategies (at various levels), PD procedures, cultural norms, and mentalities, as well as engaging in behaviors like switching from sequential to concurrent engineering and implementing elements of the lean concept in NPD, such as using set-based concurrent engineering (SBCE) rather than traditional point-based concurrent engineering (PBCE).

Front loading is described as "(i) a method that tries to enhance development performance by devoting early parts of a product development process to the discovery and resolution of difficulties." (ii) SBCE is defined by its guiding concepts rather than having a single definition. We aggregate specific definitions of SBCE in the references and reformat them as follows: In contrast to PBCE, which only iterates on one solution, SBCE is thought of as the second Toyota paradox and a method for thinking, developing, and communicating ambiguously about groups of solutions concurrently and largely independently, and making last-minute decisions in the best ongoing converging solutions. (iii) LPD is a strategy that aims to improve quality, shorten lead times, maximize value, and cut costs in PD processes [1].

Given the shortened product lifecycles, many businesses see the decrease in PD lead time as a critical issue. In the first part of the 1980s, practical designers learned about the idea of "overlapping" in the context of two interdependent PD processes, where the downstream process is started before the upstream one is finished. Overlapping has gained widespread recognition as a successful strategy for shortening the lead time for PD, largely thanks to empirical study into the Japanese manufacturing sector. It concentrated on the dissemination of early information from upstream to downstream operations during the overlapping period in a comparison of Japanese, European/American, and Japanese automotive makers. Their study encouraged numerous further investigations into the topic by stating its efficacy. However, preliminary data in the event of overlap are subject to change, and some academics have highlighted the risk of redesigning. A sequential manufacturing process, in which the upstream process is finished before beginning the downstream process, is a more suitable strategy, according to a study by Eisenhardt and Tabrizi in 1995. This is because of the rapid advancements in computer mainframe and microcomputer goods [2].

Customer Involvement

A company's ability to create new products is essential to fiscal success, and this does not happen without a customer relationship (Figure 4.4). Customers are part of new markets, and expansion opportunities are generated through innovation and the release of new products. The significance of a secure and precise NPD process is highlighted by rising global competitiveness. Companies increasingly realize how vital client pleasure is to their success. As a result, companies use various technical expertise and competencies.

FIGURE 4.4 No matter the approach, we will need to include the customer in many ways.

Understanding the requirements and desires of the consumer is necessary for market competency. The key questions are which product aspects influence consumer satisfaction and cause discontent. Satisfied customers are often dependable and create a foundation for long-term financial stability. Despite the best efforts of businesses, many NPD initiatives produce goods that fall short of consumer expectations.

As discussed, a thorough understanding of the consumer's circumstances and the ability to learn and adapt is required. Customers must actively participate in the development activity. The goal of this study was to understand the customer's needs better. To do this will require customers to be involved in the development of new products. What information do we require from the customer? When (at what stage in the development) do we need to know that? Relationship marketing is often part of the product creation process.

We have chosen to restrict the analysis to industrial manufacturing enterprises to have a concentrated focus. Additionally, since engaging individual consumers differs from engaging industrial clients and because we wanted to focus on NPD procedures in industrial relations, we eliminated enterprises that produce consumer goods (B to C). The process of developing new products in industrial companies is cross-functional. Cross-functional cooperation among the marketing, design, and production functions is necessary for new product creation. In contrast to consumer products, specialized product management/development groups are more prevalent. Cross-functional teams manage many PD processes in the business sector. This might be caused by the fact that business items are typically more sophisticated. Effective cross-functional collaboration saves costs and decreases product defects, allowing for the sale of products at a cheaper cost while maintaining a higher level of quality.

Cross-functional teams have many advantages, but it can be challenging to create effective teams because of organizational and cultural constraints. This is especially true when the team members may be distributed across the globe as modern automotive PD. Effective NPD teams may even require a skill beyond what the corporation can provide. Therefore, it may be beneficial to include individuals from outside the organization in such teams, particularly consultants and customers [3].

Market and Customer Requirements

Before producing and delivering the appropriate products, PD companies try to understand the needs of their target clients (Figure 4.5). It is common knowledge that customers are a key source of data that helps product developers minimize the risks involved in NPD projects. Several factors contribute to NPD uncertainty, but market and technology are the two biggest. An empirical study revealed that early market and technical uncertainty reduction benefited NPD project success rates. The analysis also showed that market factors, such as client needs, have a more significant impact. According to research, PD initiatives based on well-outlined customer wants (i.e., a market factor) are more likely to be successful than those based on novel technological prospects. A study by Chen et al. revealed that while market uncertainty does not alter the speed-success link, technology uncertainty does, given that speed to market is typically positively associated with NPD successes [4].

FIGURE 4.5 Regulations and standards will impact the requirements and specific design.

Similarly, it might be evident that inadequate target market understanding rather than technological understanding typically slows down or impairs development. Overall, the above research showed how crucial it is for PD projects to pay close attention to client wants compared to other considerations. Product managers, designers and developers, manufacturers, and marketers are just a few players in PD who might benefit from knowing the customer requirements. Overall, quality management aims to meet or exceed client criteria to promote customer satisfaction. This market data has historically been exploited to increase competition in the product market. Requirement elicitation, analysis, and specification are generally concerns in the management of client requirement information [5].

Connectivity

Many OEMs have telemetry systems installed on production vehicles. Key information regarding vehicle performance, environmental stimulus to which the vehicle is exposed, and use of vehicle features can be measured and off-loaded from the vehicle via the telematics systems to the OEM to what is referred to as a "back office." Consider the amount of information obtained from the volume of vehicles produced from a full production volume.

Telemetry systems are also used in the development of the vehicle. Long before a customer is able to acquire the vehicle, the OEM will be performing testing and verification activities with the vehicle. Vehicles will be driven on specialized tracks and overtime on public roads to ascertain performance. Specific performance parameters can be recorded and off-loaded for analysis.

Autonomous Vehicle

With autonomous driving systems come more sensing elements, more ECUs, and complex system interactions. Specific events and performance thresholds are identified for recording, for example, autonomous hard braking events and certain vehicle stability parameters would be interesting to understand. These events may be recorded and off-loaded via telemetry systems.

Big Data

This volume of "back office" collected data from the vehicle provides a great source of information for new products or refining or adapting existing vehicle applications. The development of the product is often based on missing or assumed concepts and ideas. The mass of data collected from the entire production volume of the OEM can refute or confirm the things we believe about the vehicle and application use.

Virtual Assistants and Robots into Daily Life: Vision and Challenges

On July 10, 2017, a suspect accused his girlfriend of cheating and threatened to kill her if she called the police. Alexa, Amazon's virtual assistant, may have saved the victim's life by dialing 911. This incident is a sign of the increasing pervasiveness and power that artificial intelligence (AI) will have in our daily lives, even though the gadget may have misconstrued the suspect's question to his girlfriend, "Did you contact the sheriff?" as a request to dial 911. The idea of robot helpers has existed since antiquity.

According to Greek myth, Talos, a robotic statue, was created by Hephaestus, the god of invention and technology. Talos was created to protect Europa at Zeus's request, just as Alexa may have protected the suspect's girlfriend by dialing 911. Although the idea of an automaton predates civilization, AI has only recently started to be defined and implemented in real terms. The definition of cybernetic as "the scientific study of control and communication in both living beings and machines" was established by Norbert Wiener in 1948 [6].

After being challenged, this definition is now used to characterize several computer science fields, including AI and robotics.

To determine whether machines can accurately mimic human behavior and, thus, actually think, Alan Turing developed the "Imitation Game" experiment in the 1950s. By the year 2000, according to Turing, such machines would be commonplace. Many may believe that this is a very doable task. However, one of the earliest objectives in computer science is the imitation of human behavior. AI is defined as "the ability of a machine to simulate intelligent human behavior" using these definitions.

Cybernetic applications, such as virtual assistants, chatbots, and social robots, are coming under more and more attention thanks to developments in low-cost manufacturing sensors, algorithmic progress, networking capabilities, and shrinking hardware.

Many AI applications imitate one or more aspects of human cognition, most notably naturalistic dialogue. Companies like Amazon, Apple, Google, Samsung, and others are quickly adopting natural language interfaces (NLI), and it has been projected that the next significant change in user experience designs will be in NLI, including conversational agents in robots and virtual assistants [7].

Six Sigma and Lean Manufacturing Techniques

According to empirically supported research, a structured approach to process improvement is more productive. Methodologies like Six Sigma are collections of tasks carried out in a specific order to achieve predetermined values (Figure 4.6). Tools are defined as devices that make statistical or other types of data analysis easier.

FIGURE 4.6 Design for Six Sigma approaches applies to the design of the product.

Wright Studio/Shutterstock.com.

Six Sigma is defined at Allied Signal, a business that successfully adopted this methodology, as a paradigm shift for accelerating the rate of process and product improvements. Six Sigma is a group of process improvement methods used methodically in several projects to reach high levels of reliability. It is founded on standards established by quality experts. The Greek letter (σ) that is typically used to refer to the standard deviation, a measurement of the variation or dispersion in a process output around its mean value (μ), makes up the phrase "Sigma." Six Sigma quality quantitatively indicates that less than two defects per million opportunities depart from the upper and lower specification limits.

This level has nearly no flaws. Operating at a low defect level in some industries could not be cost-effective. However, achieving a high defect-free level is crucial for high-yield companies like Motorola that manufacture electronic parts with thousands of potential failure modes due to the various parts used in every product to keep the total opportunity for failure as low as possible [8]. These electronic parts will end up on vehicles. Additionally, critical vehicle systems may have performance expectations that will require a Six Sigma approach (Figure 4.7).

FIGURE 4.7 Cost of poor and good quality curve example.

© SAE International.

Lean Manufacturing Tool to Enhance Productivity in Manufacturing

Any company wants to satisfy its customers, and this may be done by providing a high-quality product on schedule for a fair price (Figure 4.8). However, suppose an organization is adaptable enough to consistently and systematically respond to consumer needs and add value to the product in the process. In that case, it will survive and maintain its competency and market share regardless of whether it provides manufactured products or services.

Artic_photo/Shutterstock.com.

The main factors determining a product price are labor, material, and equipment costs, which rise together with the inflation rate. According to basic math, a direct financial cost is associated with underusing the tools, and resources. Labor or talent is a little more complicated; we need to engage our team, which does not necessarily mean 100% of the hours are applied to a defined effort. Therefore, maximizing these main parameters must come first, lowering manufacturing activity waste. All manufacturing industries rely heavily on lean manufacturing techniques, including automotive, electronics, plastic, textile, food, dairy, foundry, stamping, and maintenance.

Reduction in cycle time, removal of non-valued operations, and a clean, orderly, and hygienic workplace were the benefits noticed following the deployment of an individual or combination of lean manufacturing techniques. In addition, there will be a smooth production flow, a productivity improvement, a decrease in production costs, employee involvement, order recording, inventory reduction, breakdown, and increased intra- and interconnection for swift decision-making [9].

Understanding Customer Needs and Wants

The marketing management process oversees recognizing, foreseeing, and profitably addressing client needs. The responsibility for connecting and organizing operational activities and coordinating the efforts of all commercial areas falls on the marketing managers. They must also collaborate closely with other departments (such as finance, maintenance, operations, etc.) to add value for both the client and the company.

Marketing managers must consider the true worth of their goods and services to the public. Due to the interdependence of service production and consumption, it has been proposed that marketing for travel and tourism entails co-creating value since ethereal, dynamic resources are more significant than tangible, static resources. Therefore, to ensure their long-term value throughout time, marketers of today are expected to anticipate the demands and wishes of their specific customers. Naturally, this calls for an in-depth market study and analysis. As a result, it is crucial that businesses go through a thorough consumer and market study process. It is in their best interests to research the current market, consumer needs and wants, and potential solutions.

Market research is vital to look into potential competition and forecast market developments. The systematic planning, gathering, analysis, and reporting of data and findings pertinent to a particular marketing issue a company is experiencing may be referred to as market research. They may make better decisions on creating a product or service after processing and analyzing the data gathered through market research, which gives them the necessary knowledge base. As a result, the knowledge produced by marketing research is divided into the categories of making, enabling, and fulfilling a promise [10].

Virtual and Augmented Reality Technologies for Product Realization

One of the most significant obstacles facing engineers, who produce most of the wealth in our society, is the realization of products (Figure 4.9). The complexity of product realization activities, in terms of their overall quality, costs, and lead time, is rapidly increasing as society advances. In the meantime, our society has developed into a more holistic and lifecycle-oriented attitude toward creating, using, and recycling engineered items environmentally friendly.

FIGURE 4.9   Virtual reality (VR) and augmented reality (AR) help develop the product and the manufacturing line.

Gorodenkoff/Shutterstock.com.

Seeking out improved approaches and technologies to support product realization in the engineering profession has been increasing with these new societal developments. As of now, digital computers are a necessary instrument for product realization. Computers are used to improve quality, lower costs, and minimize lead times. In addition, engineers employ computers to support decision-making and process control during the design, planning, production, and distribution phases. As a result, how engineers and computers communicate now plays a crucial role in determining how successful products are developed overall. Current computer-aided engineering technologies provide users with two-dimensional (2-D) and three-dimensional (3-D) graphical and textural interfaces. Recent advancements in virtual reality (VR) and augmented reality (AR) technology, as well as multimedia (2-D), offer some extremely intriguing potential to satisfy this urgent demand [11].

Configuration Management Is a Key Product Management Area in 3-D Printing

3-D printers allow engineers to quickly explore the physical attributes of the product (Figure 4.10). In the past, parts were built off prototype tools. These parts would be given to the customer as exploratory products, which would take time and be costly. These parts we hand to the customer to get feedback. Changes can be made quickly to the computer design with another product reprint for exploration.

FIGURE 4.10   Configuration management is a key product management area in 3-D printing.

master_art/Shutterstock.com.

Product Performance, Function, and Physical Attributes

Physical and functional configuration management (CM) are the two fundamental categories. The conformity of the product to its defining documents is evaluated through physical verification. Functional verification determines the compliance of a product with all specified applicable requirements. Depending on the requirements of the program or the client, physical and functional attributes may be required to effectively manage the development of the product. This will result in incremental tests or inspections, or they may be the project and product audits topic.

The discipline of CM is used to manage the technical direction of the product. CM provides a roadmap in the development of the product and what physical and functional attributes are next in the product. It also explains how the product has made it to this specific point. Tools such as surveillance audits identify and record the functional and physical characteristics of a product end item, control changes to those characteristics, record and report change processing and implementation status, and verify compliance with specific requirements.

From the standpoint of a product, CM is a management procedure for developing and preserving consistency among the performance, function, and physical characteristics of a product with its requirements, design, and operational data throughout its life. The accurate, current configuration of defense assets and their connections to related documents are made known by CM. The CM process effectively handles required modifications, ensuring that all operation and support effects are considered. Although they ought to be evident, the advantages of the method are frequently disregarded.

CM is an integrated and well-documented set of management controls used through production and post-production product support (Figure 4.11). It is intended to assist in converting requirements into products that will perform as required and that can be produced, operated, and supported as planned. It supports and helps the program or system deliver products faster and more efficiently [12].

FIGURE 4.11 CM ensures control of the product branches and remerging.

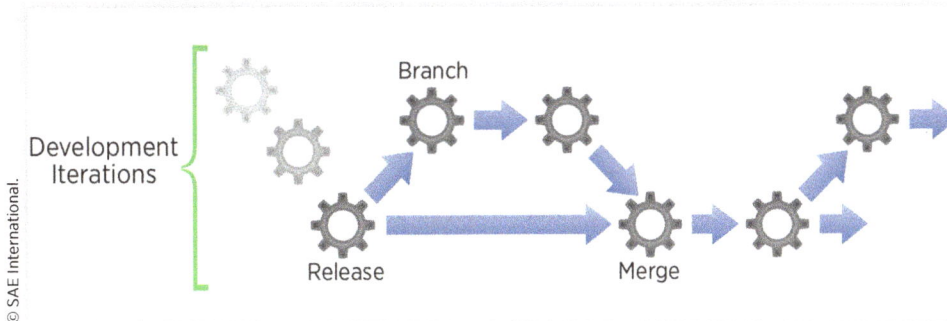

In the course of the development of the product and after the product launch, we may modify the product. There may be more than one active version of the product at any given time, including the development versions. There are competitive reasons for converging the version back into one.

References

1. Belay, A.M., Welo, T., and Helo, P., "Approaching Lean Product Development Using System Dynamics: Investigating Front-Load Effects," *Advances in Manufacturing* 2, no. 2 (2014): 130-140, doi:10.1007/s40436-014-0079-9.

2. Itohisa, M., "Overlapping and Frontloading Seeking an Effective Product Development Process," *Annals of Business Administrative Science* 12, no. 6 (2013): 291-309.

3. Lagrosen, S., "Customer Involvement in New Product Development: A Relationship Marketing Perspective," *European Journal of Innovation Management* 8, no. 4 (2005): 424-436, doi:https://doi.org/10.1108/14601060510627803.

4. Chen, J., Reilly, R., and Lynn, G., "The Impacts of Speed-to-Market on New Product Success," *IEEE Transactions on Engineering Management* 52 (2005): 199-212, doi:10.1109/TEM.2005.844926.

5. Chong, Y.T. and Chen, C.-H., "Customer Needs as Moving Targets of Product Development: A Review," *International Journal of Advanced Manufacturing Technology* 48, no. 1 (2010): 395-406.

6. Wiener, N., *Cybernetics, or, Control and Communication in the Animal and in the Machine* (New York: Wiley, 1948).

7. Rawassizadeh, R., Sen, T., Kim, S.J., Meurisch, C. et al., "Manifestation of Virtual Assistants and Robots into Daily Life: Vision and Challenges," *CCF Transactions on Pervasive Computing and Interaction* 1, no. 3 (2019): 163-174.

8. Salah, S., Carretero, J.A., and Rahim, A., "Six Sigma and Total Quality Management (TQM): Similarities, Differences and Relationship," *International Journal of Six Sigma and Competitive Advantage* 5, no. 3 (2009): 237-250.

9. Palange, A. and Dhatrak, P., "Lean Manufacturing a Vital Tool to Enhance Productivity in Manufacturing," *Materials Today: Proceedings: Part 1* 46 (2021): 729-736.

10. Camilleri, M.A., "Understanding Customer Needs and Wants," in *Travel Marketing, Tourism Economics and the Airline Product* (New York: Springer, 2018), 29-50.

11. Lu, S.-Y., Shpitalni, M., and Gadh, R., "Virtual and Augmented Reality Technologies for Product Realization," *CIRP Annals—Manufacturing Technology* 48, no. 2 (1999): 471-495.

12. Sheng, R., "Chapter 6: Managing People, Product, and Process (P3) Implementation," in Sheng, R. (Ed.), *Systems Engineering for Aerospace* (London, UK: Academic Press, 2019), 75-112.

5

Readiness Level

Maturity

This chapter describes the concept of maturity, technology readiness level (TRL), manufacturing readiness levels (MRLs), and actions we take to confirm that level and advance us to the next level of maturity (Figure 5.1). Maturity in this context has an analog we all understand: children, adolescents, and adults. In this case, we are describing technology and manufacturing.

FIGURE 5.1   Readiness levels for design and manufacturing.

It is probably easy to see how the product matures. The prototype parts throughout the development effort have maturity levels, for example, the first level of prototypes is often a mock-up, with no features but an exploration of the physicals, geometries, and dimensional attributes of the product. The next level can be some features and refinement of the physicals, probably of some soft tool. These parts have some level of capabilities, and the customer may be able to use these parts to explore. In the last level, the product has complete feature content and is likely off the production tools and processes. There is so much that happens before the customer volume production.

Technology

Technology is a set of scientific ideas and knowledge that humans use to attain a specific goal, which can be the solution to one particular problem of the individual or the satisfaction of some of his needs (Figure 5.2).

FIGURE 5.2   Technology and product development are parallel with manufacturing development.

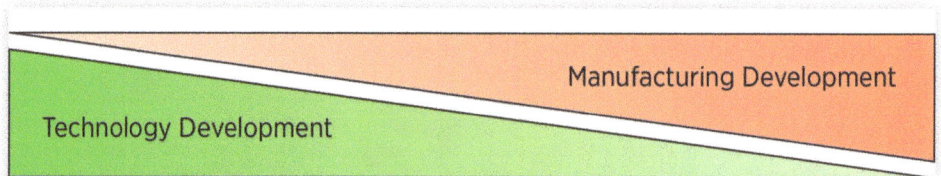

Technology Development

Manufacturing Development

It is a broad concept encompassing various aspects and disciplines within electronics, art, and medicine. For example, robots are created to automate repetitive tasks or animal cloning.

Types of Technology

Technology comes to have different meanings, but they all apply to the same central idea: to perform a type of task that helps people modify their environment to meet a need. It consists of different knowledge, tools, techniques, and devices. In this way, technology has been devised by people to be able to solve those jobs that they need to solve.

They have a clear goal and use precise tools. Their uses may be profound or minor, but they are intended to have a clear purpose in all cases. Each end can have a technological idea, a tool, or a function that deals with achieving it. This leads to, as we said, technology being classified depending on the parameter in various ways. Thus, we can distinguish different types of technology depending on their applications or purposes:

- Clean Technology and Materials
- Soft and Hard Technologies
- Heart Rate Acceleration
- Flexible and Fixed Technology
- Technology of Operation, Equipment, and Product

The file titled "Capturing Design and Manufacturing Knowledge Early Improves Acquisition Outcomes" the report to the subcommittee on readiness and management support, committee on armed services (2002) [1]. United States Department of Defense DOD has historically developed new weapon systems in a highly concurrent environment that usually forces acquisition programs to manage technology, design, and manufacturing risk at the same time. This environment has made it difficult for either DOD or congressional decision makers to make informed decisions because appropriate knowledge has not been available at key decision points in product development. DOD's common practice for managing this environment has been to create aggressive risk reduction efforts in its programs. Cost reduction initiatives that typically arise after a program is experiencing problems are common tools used to manage these risks.

Concept Decision Matrix

A decision matrix is a technique for assessing options and choosing the best prospect among alternatives. It is a handy tool when you have to decide between more than one option, and there are several factors that you need to consider to make the final decision. You have probably heard about the decision matrix but with another name, although, in reality, it is talking about the same thing. Among those names are the following:

- Pugh Matrix (Figure 5.3)
- Grid Analysis
- Multi-attribute Utility Theory
- Problem Selection Matrix
- Decision Grid

FIGURE 5.3 Example of a Pugh Matrix.

Pugh Matrix   2005-08-22
Function name: New transmission shifter

Evaluation criteria	Priority— 5 is high	Shifter A		Shifter B		Knob shifter dash		T-handle on dash		Shifter on column	
		Rating	Weighted	Rating	Weighted	Rating	Weighted	Rating	Weighted	Rating	Weighted
Product cost	5	1	5	3	15	2	10	2	10	1	5
Tooling cost	2	5	10	4	8	2	4	3	6	1	2
Development time	3	4	12	2	6	1	3	1	3	1	3
Ease of entrance into sleeper compartment	5	1	5	1	5	3	15	3	15	3	15
Sight location of shifter during driving	2	0	0	1	2	3	6	2	4	3	6
Issues with extremely large people	4	1	4	1	4	4	16	4	16	1	4
Conventional acceptance	5	3	15	4	20	2	10	3	15	1	5
Instrument panel space	3	4	12	4	12	1	3	1	3	4	12
New steering wheel with up/down and man/auto buttons	1	2	2	2	2	1	1	2	2	2	2
Reach for manual shifting	3	2	6	3	9	4	12	1	3	2	6
Clinic rating	5	3	15	3	15	3	15	3	15	1	5
Ease of diagnostics	3	2	6	2	6	1	3	1	3	1	3
Number of new parts	1	2	2	2	2	1	1	1	1	1	1
Total score			94		106		99		96		69
Selected/Rejected/ Maintained alternative											

Responsible for Pugh

Leader: TL	Advanced Engineering: Jer	Product Planning: MM
Engineering: BT	Aftermarket: Jep	Marketing: JF
Engineering: JY	Product Planning: BH	Eng. PM: RK

When to Use the Decision Matrix

The process is solid and relatively straightforward, but it is most effective when deciding between several comparable options. If the evaluation criteria are not the same for the different options, then it is very likely that the matrix is not the best decision-making tool. For example, a decision matrix will not help you decide what direction your team should take for the coming year because the options you must choose are not comparable.

Use the Decision Matrix for the Following Cases

- Compare several similar options.

- Help narrow down the number of options until you reach a final decision.

- Weigh between a variety of important factors.

- Approach the decision from a logical point of view instead of an intuitive or emotional one.

Manufacturing Reviews

The pharmaceutical industry has considerably intensified quality and regulatory requirements in recent years. This rigor in the production of drugs has made it possible to achieve mastery of pharmaceutical processes. A summary report is carried out annually for each product, called the annual product quality review. It overviews all the events impacting a product over a year. The yearly product quality review is much more than a simple response to regulatory requirements but a major process in improving product quality. It is a tool for monitoring the quality of products over the years. It ensures control of the manufacturing process by following trends, anticipating drifts, evaluating the specifications, method of manufacture, and control of the product (Figure 5.4). First, the notion of quality on a pharmaceutical production site and the regulatory requirements of the annual product quality review and its characteristics will be presented. Second, a demonstration of the annual quality review as a quality management and improvement tool will be developed. Finally, the latest indicators proposed by the Food and Drug Administration (FDA) will be highlighted.

FIGURE 5.4 Level of automation will be one of the things we will consider in the manufacture of the product.

August Phunitiphat/Shutterstock.com.

According to research by D. Wheeler and M. Ulsh DOE can use Manufacturing Readiness Levels (MRLs) to address the economic and institutional risks associated with a ramp-up in polymer electrolyte membrane (PEM) production. "Investment risk of developing manufacturing capability for hydrogen and fuel cell technologies is high" [2].

According to Interim rule, Federal Register (June 2011), DOD is issuing an interim rule to implement section 812 of the National Defense Authorization Act for Fiscal Year 2011. Section 812(b)(5) instructs DOD to issue guidance that, at a minimum, shall require appropriate consideration of the manufacturing readiness and manufacturing-readiness processes of potential contractors and subcontractors as a part of the source selection process for major defense acquisition programs [3].

Manufacturing Risk Analysis

Risk analysis is a method of identifying and quantifying the likelihood of a product failure (Figure 5.5). Risk analysis is an essential component of the design for manufacturing (DFM) process, which includes design for assembly (DFA) and failure mode and effects analysis (FMEA). It provides the opportunity to identify risks early in the manufacturing stages prior to series production to avoid significant impact to long-term production volumes. Companies that have successfully implemented risk analysis in their development cycle have reported strong benefits to customer satisfaction and reduced engineering (R&D) expenses.

FIGURE 5.5 Reviews of documents, drawings, and other artifacts are important communications mechanisms.

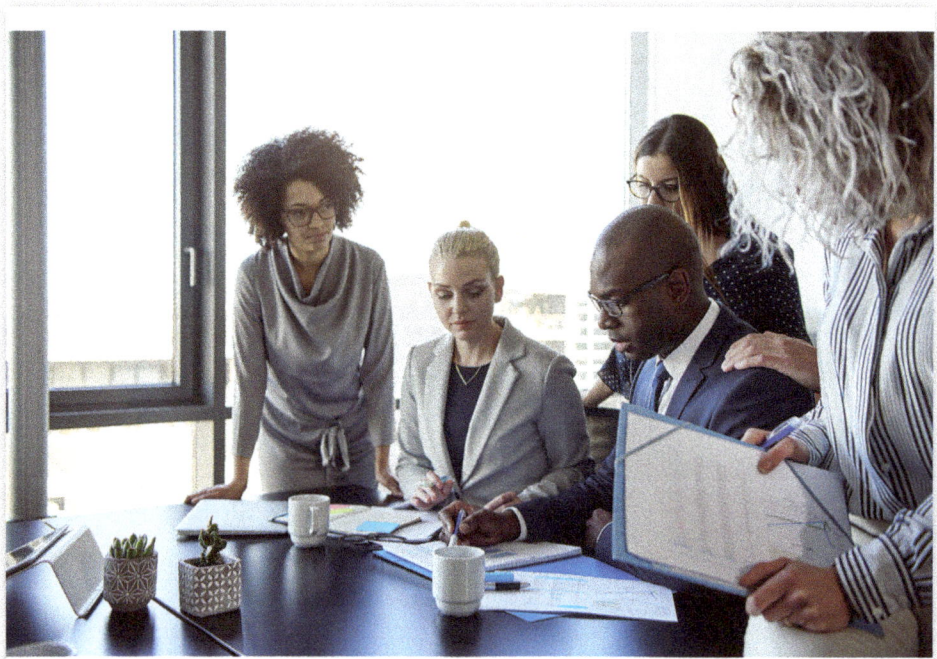

Ground Picture/Shutterstock.com.

Design for Assembly

Confirming product manufacturability and assembly prior to product launch is a critical aspect of product development processes. This is also looked upon as an opportunity to improve the manufacturing process and identify defects in the early stages. DFA is a great way for many companies to develop high-quality products in less time at lower production costs. Higher quality at a lower cost usually means more sales and great customer loyalty.

DFA also enables improving the product portfolio by harmonizing the part or component structure focusing on

- Part number reduction, thereby making manufacturing and product assembly easier
- Better handling of parts during assembly
- Lower maintenance of variability
- Improved product quality

The figure provides a good overview of how the design can influence production costs and, by reducing part count, improve DFM and optimize production processes. In addition, the use of DFA on cost savings is exponential with the part count reduction (Figure 5.6).

FIGURE 5.6 DFM and DFA.

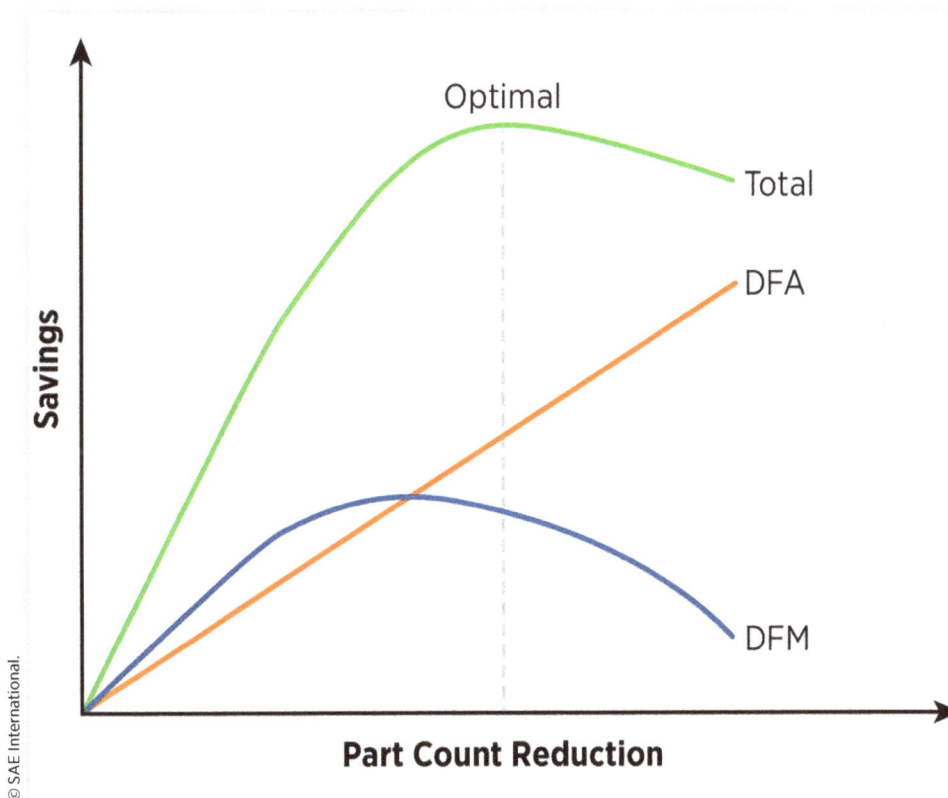

Failure Mode and Effects Analysis

The FMEA is a method used to prevent failures and analyze the risks of a process by identifying causes and effects to determine the actions that will be used to inhibit failures and their use in Machine Learning applied to production processes (Figure 5.7).

FIGURE 5.7   FMEA will be part of testing the design and the manufacturing proposal.

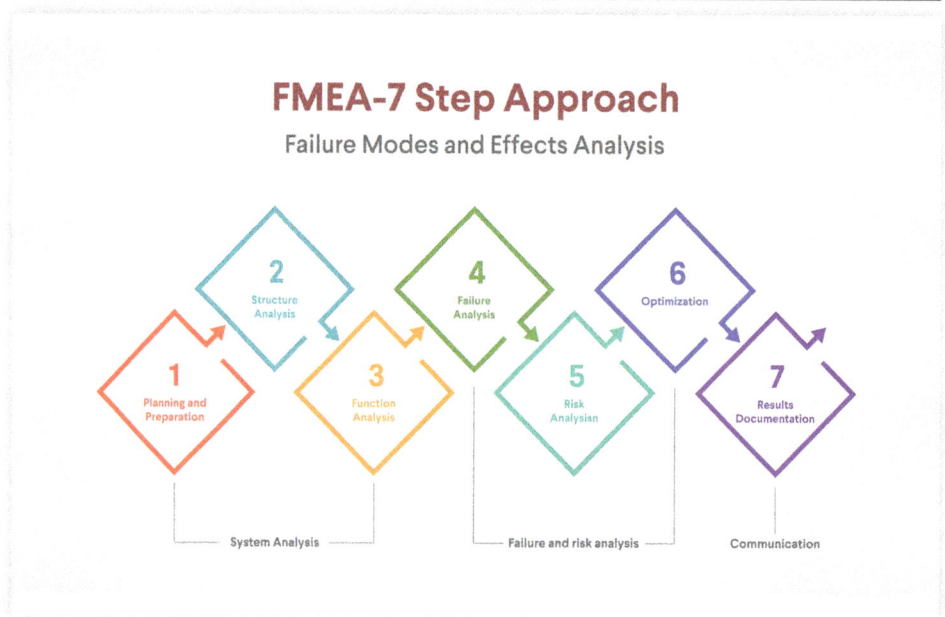

The failure mode is connected to how a process might be caused to fail and comprises three elements: effect, cause, and detection. The effect is the consequence of what the failure can cause to the customer; the cause indicates the reason why the error occurred, and the detection is the way used to control the process to avoid possible failures.

The FMEA aims to identify, delimit, and describe the nonconformities (failure mode) generated by the process and its effects and causes through preventive actions to reduce or eliminate them.

FMEA Types

- **Product FMEA:** It is related to the failures that can occur in the product within the product and project specifications.

- **Process FMEA:** Related to the failures that may occur in the process planning, process controls, and material handling and are considering the nonconformities presented in the product related to product manufacturing and project specifications.

Manufacturing 4.0

Manufacturing 4.0 refers to a new way of producing by adopting 4.0 technologies—solutions focused on interconnectivity, automation, and real-time data (Figure 5.8).

FIGURE 5.8 Manufacturing 4.0 considers the production, team member involvement, as well as the data generated and used.

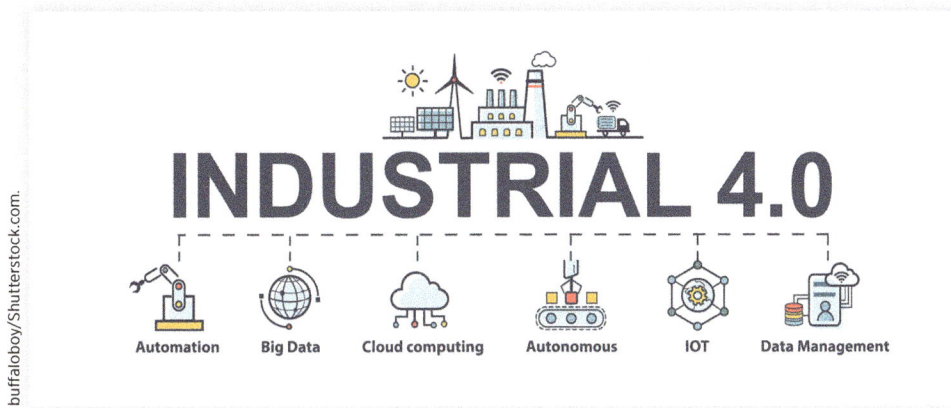

This transformation covers the production of your company's goods and/or services and the entire value chain since it reconfigures production processes and product services, business management, customer and supplier relationships, and, in a broader sense, business models.

To transform into a Manufacturing 4.0, your company must gradually incorporate different innovative technological components from the digital and physical domains. **For example**,

- AI
- Internet of things
- Robotics
- 3-D printing
- Cloud services
- Cybersecurity

For this, you must identify what type of technology is the most functional for your company so that it is beneficial to make a significant investment. **Consider the following**:

- What is the good or service that your company provides?
- What business model do you adopt?
- What is your degree of process automation?
- How is production adapted to the needs of the company's staff?
- What level of interaction exists between suppliers and customers?

Data-Driven

One of the main drivers of Manufacturing 4.0 is data integration. Shop floor computers have been developed to promote cooperation and creativity by broadening the scope of data collecting and making information conveniently accessible.

Today, shop floor computers are expanding the scope of data capture to include detailed information on all elements of each specific step of the production process, including manufacturing and monitoring, to make that information available to end users.

What Is Data Integration in Manufacturing? Data integration can be defined as the ability to mesh information technology (IT) with operations technology so that manufacturers can fully harness the value of the data generated by their factories.

In other words, it enables the free flow of information through every stage of the manufacturing cycle by interconnecting digital and physical systems across all manufacturing operations.

This is achieved by connecting the main systems through an integrated enterprise resource planning (ERP), a manufacturing execution system (MES), and supervisory control and data acquisition (SCADA) system to better understand the retrieved data.

How Data Integration Works in the Industry ERP, MES, and SCADA systems contain highly valuable data for maximizing manufacturing efficiency and quality.

The data integration process in the production plant is carried out within the levels of the industrial automation pyramid with the interconnection of the ERP and MES systems to the SCADA system that captures the information through sensors controlled by PLCs.

In short, ERP systems inform manufacturers of inventory levels and delivery times; MESs track and manage manufacturing information in real time, providing golden insights on traceability and performance, and SCADA systems are used in the monitoring and control of industrial equipment at various stages such as development, manufacturing, and production.

Connectivity

Manufacturing connectivity was born in the light of the Fourth Industrial Revolution, also called intelligent manufacturing; it is a process in practically the whole world aimed at increasingly digitizing the industrial field—talking about the new Manufacturing 4.0 (Figure 5.9).

FIGURE 5.9   Connectivity throughout the enterprise reduces cost and risk.

elenabsl/Shutterstock.com.

It consists of the digitization of machinery and production processes connected to each other within a company, using the Internet and technologies that simplify and digitize the handling of machinery.

To achieve this, there are many tools on the market, including wireless equipment, GSM and GPRS networks, connectors for industrial communication, and innovative technologies that allow those responsible for the processes to monitor, manage, and program functions of their machinery, either in person or remotely.

With such technological tools, companies can now connect, for example, the production and supply areas, among other things. Digitizing industrial machines is the beginning of the transformation process towards Manufacturing 4.0.

Maturity Level

We can define various levels of maturity of increasing complexity in additive manufacturing. It is the translation of the steps usually taken by companies in the use of additive manufacturing and entails a series of actions to go from one level to another.

The sequence and intensity of these actions will define the implementation strategy of each company. There are five levels, which we detail below:

Level 0: Preproduction At this level, only additive manufacturing for prototypes is contemplated. It is used in:

- Concept development
- Prototypes
- Display
- Assembly process validation

Level 1: Direct Applications Indirect use is made in production processes by applying technology in tools, fixtures, or molds. For example:

- Welding jigs
- Tools and terminals for assembly robots
- Soft collets for CNC
- Inspection accessories
- SLA patterns or models
- Cores for sand molds
- Resin or silicone molds
- Molds for the manufacture of composite materials: soluble males, tools, and molds for laminating

Level 2: Parts Replacement At this level, the use in the final part is already contemplated but in a piece-by-piece substitution model, for example, for the manufacture of spare parts, so that the design or functionality of the initial part is not modified and complies with the original requirements.

It involves introducing additive manufacturing in the supply chain to reduce out-of-stock risk, logistics optimization, reduction of supply or inventory times, and warehouse expenses.

Level 3: Functional Parts Level 3 contemplates using techniques that improve the behavior or functionality of the parts, such as topological optimization or the integration of subassemblies, to take advantage of the additional benefits of additive technology (e.g., APWORKS Light Rider).

Level 4: Functional Integration This level is of advanced use, in which products have been adapted for maximum multifunctional efficiency thanks to additive manufacturing. Processes have been adapted to fully integrate workflow and digital flow into the company's tools.

Performance optimization and customization possibilities are fully exploited. Likewise, several design iterations are applied to optimize some of the characteristics of the final piece (weight, heat transmission, rigidity, etc.), in addition to integrating various components in use or customizing functionality (e.g., hydraulic manifold optimized in weight and fluid dynamics, from AIDRO).

Maturity Models

A maturity model offers different maturity levels describing the stages through which systems should evolve. This model allows companies to identify their strengths and weaknesses regarding key characteristics: interoperability, security, agility, etc. The goal is to predict and/or diagnose potential problems and prioritize the actions needed to resolve them.

The degree to which an organization, or an organizational unit, develops, assimilates, and implements good practices in managing projects, programs, and portfolios is known as project management/management maturity. Maturity models may be used to assess an organization's or organizational unit's level of project management maturity.

A maturity model is a well-structured set of elements (good practices, measurement tools, analysis criteria, etc.) which allow for identifying the capabilities installed in project management in the organization, comparing them with standards, identifying gaps or weaknesses, and establishing continuous improvement processes.

An example of maturity models in project management can be found from the Capability Maturity Model Integration (CMMI) developed by the Software Engineering Institute, SEI, at the request of the Federal Government of the United States (US) in 1986 for the evaluation of processes related to software development. The objective of this model was to provide a questionnaire that would serve as a tool to identify the areas where the software development processes needed improvement (Figure 5.10).

FIGURE 5.10 Maturity levels.

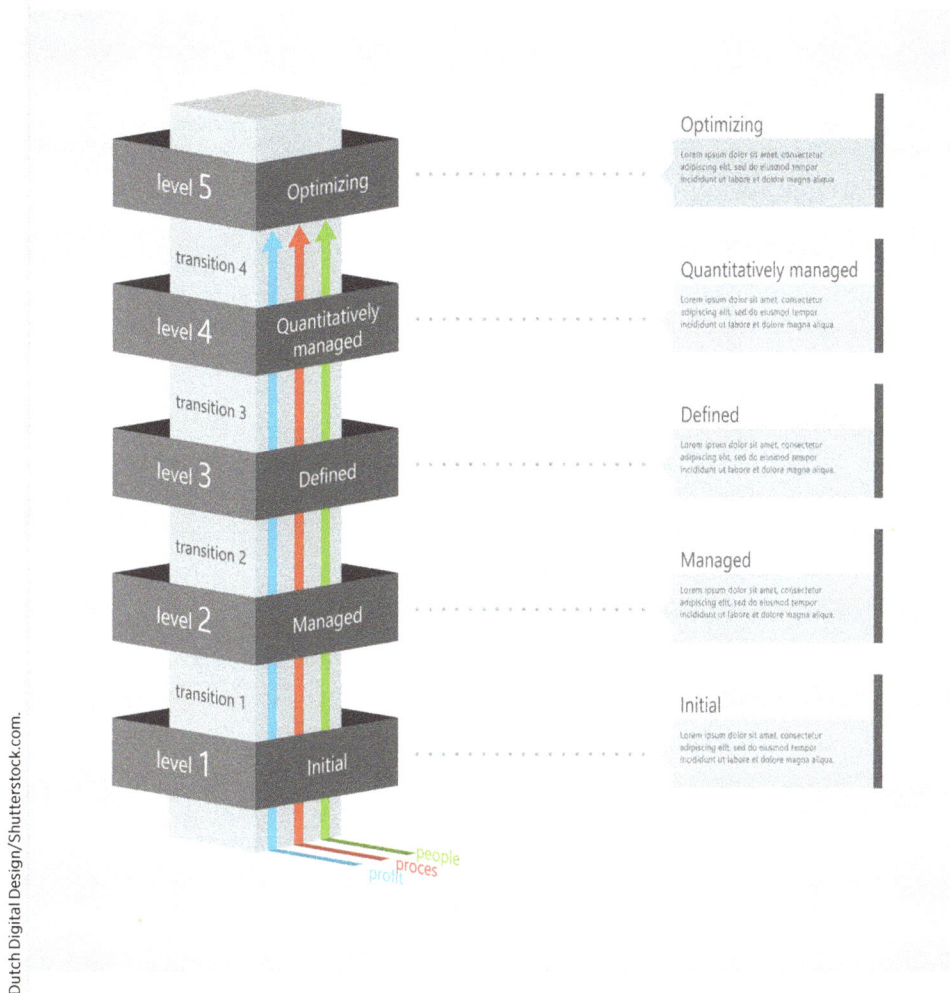

Capability Maturity Model Integration

The CMMI is a methodology used to develop and refine the maturity of an organization's software development process. SEI developed the said methodology in the mid-1980s. It is a process improvement approach.

Evaluate an organization against a scale of five levels of process maturity. It is about what processes should be implemented and not so much about how the processes should be implemented. Each maturity level consists of a specified set of process areas known as key process area (KDA), which include the following KDA: Goals, Commitment, Capability, Measurement, and Verification.

The CMMI levels are as follows:

- **Level One: Initial**. The work is done informally.
  Ad hoc activities describe a software development organization at this level (the organization does not plan).

- **Level Two: Repeatable**. Work is planned and tracked.
  This level of software development organization has basic and consistent project management processes to follow cost, schedule, and functionality. The process is underway to replicate previous successes in projects with similar applications.

- **Level Three: Defined**. The work is well defined.
  At this level, software processes for management and engineering activities are defined and documented.

- **Level Four: Managed**. The work is quantitatively controlled.
  Software Quality Management: Management can effectively control software development through accurate measurements. At this level, the organization established a quantitative quality goal for the software process and maintenance.
  Quantitative Process Management: At this level of maturity, process performance is controlled using statistical and other quantitative techniques and is quantitatively predictable.

- **Level Five: Optimization**. The work is based on continuous improvement. The key characteristic of this level is to focus on continuously improving the process.

The Key Features are:

- Process change management

- Technological change management

- Defect prevention

TMMi

The Test Maturity Model Integration (TMMi) framework was developed by the European organization TMMi Foundation and originated in the Test Maturity Model (TMM) project developed by the Illinois Institute of Technology. TMMi, based on CMMI, has been developed as a staged model, using predefined sets of process areas so that an improvement path for the organization can be defined through a maturity model.

The TMMi maturity levels are

- **Level 1: Initial**
  Represents a condition in which there is no formally documented or structured testing process. Tests are typically developed on an ad hoc basis after code is written, and testing is treated the same way as debugging. The goal of testing is to prove that the software works. It depends on "heroes," and there is no understanding of the cost of quality.

- **Level 2: Managed**
  The second level is reached when testing processes are separated from debugging. It can be achieved by establishing test policies and objectives, introducing the steps in a fundamental test process (e.g., test planning), and applying basic test methods and techniques. Test environments are used.

- **Level 3: Defined**
  When a testing procedure is integrated into the software development lifecycle, the third stage is attained and documented using formal standards, procedures, and

methods. Reviews are performed, and there should be a separate software testing function that can be controlled and monitored. Non-functional tests are performed.

- **Level 4: Measured**
  Level four is achieved when the testing process can be effectively measured and managed at the organization level to benefit specific projects. A product quality evaluation process is implemented.

- **Level 5: Optimized**
  The last level represents a state of test process maturity where data from the test process can be used to help prevent defects, and the focus is on optimizing the established process. A permanent test process improvement group is established.

TMMi does not have a separate process area for test tools and/or test automation. Within TMMi, test tools are treated as a support resource (practices) and are, therefore, part of the process area they support. For example, the application of a test design tool is a practice of support tests in the process area of design and execution of tests at the TMMi 2 level, and the application of a performance test as a tool is a practice of support tests in the process area of Non-functional Tests at TMMi 3 level.

Reasons for TRL

Like any assessment tool, this technology readiness indicator has advantages and limitations. Among its advantages, we can mention the common vision of the state of the art of technology, the risk management, the usefulness for making decisions concerning the financing of technology, or the representation of the technical efforts still necessary. But the TRL scale also has limitations. Thus, a mature technology may not apply to a need, while another less mature one may be better suited. Indeed, many factors can come into play (cost, market share, competition from other technologies, environmental impact, etc.) that the relatively simple concept of TRL (single rating) only sometimes makes it possible to assess properly.

Gain Confidence in the Design Maturity

TRL is about assessing the quality of a product design and understanding the requirements of a product design. It helps formulate the product design strategy and ensure that it evolves to meet its users' needs. This helps ensure that the product meets all its users' requirements and provides them with value for money.

When creating a new product, it can take a lot of work to know where to start. You may need to check if you need a new feature or just minor tweaks. With TRL, you can assess your product design quality and identify areas that need improvement before developing.

Provide Early Product V&V

In the past, the notion of V&V was largely a function of the product itself (Figure 5.11). It was assumed that its creation was complete once a new product or service was created. The V&V stage would come later after products were implemented in the market and customers had begun to use them.

FIGURE 5.11   V Model specification and verification.

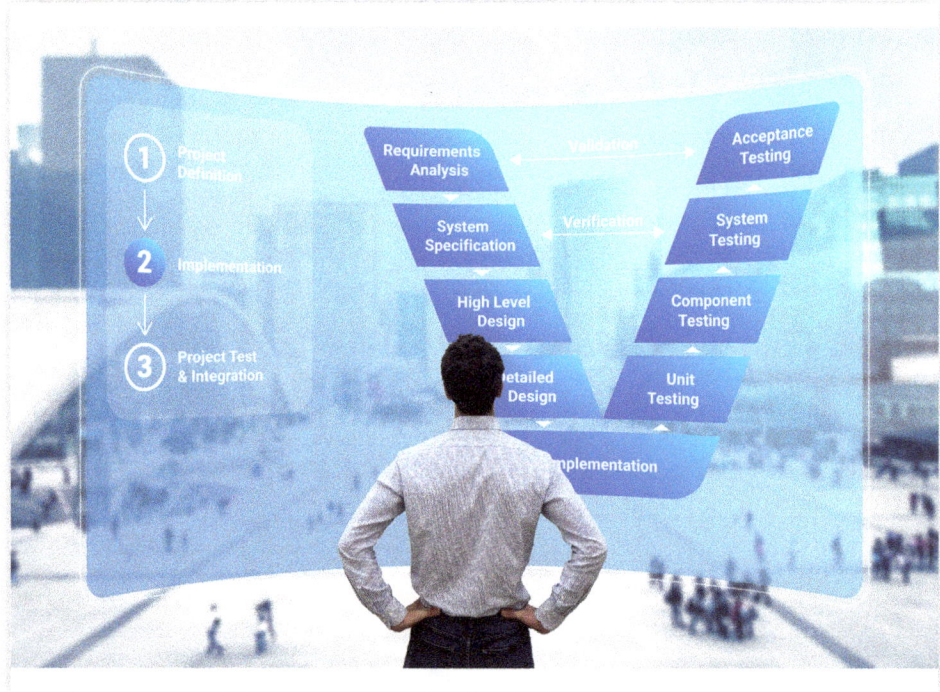

This program is meant to help companies get their products out to the market faster and improve the quality of their products. TRL helps manufacturers avoid having their products fail in the marketplace, resulting in significant financial losses.

The purpose of TRL is to provide early product V&V. Because they are developed before mass production begins, TRLs can be used by engineers as a way of testing parts or even entire designs before production begins.

Checkpoint for Investment and Tooling Readiness

The purpose of the TRL is to check for investment and tooling readiness. The degree to which it is completed can affect the quality of a product and how long it takes to develop. TRL is a tool to determine whether your product is ready for investment and tooling. It is based on the concept of a checkpoint, which is used to determine whether a product can be shipped (Figure 5.12).

FIGURE 5.12 Maturing an electronic control unit requires building hardware and software iterations.

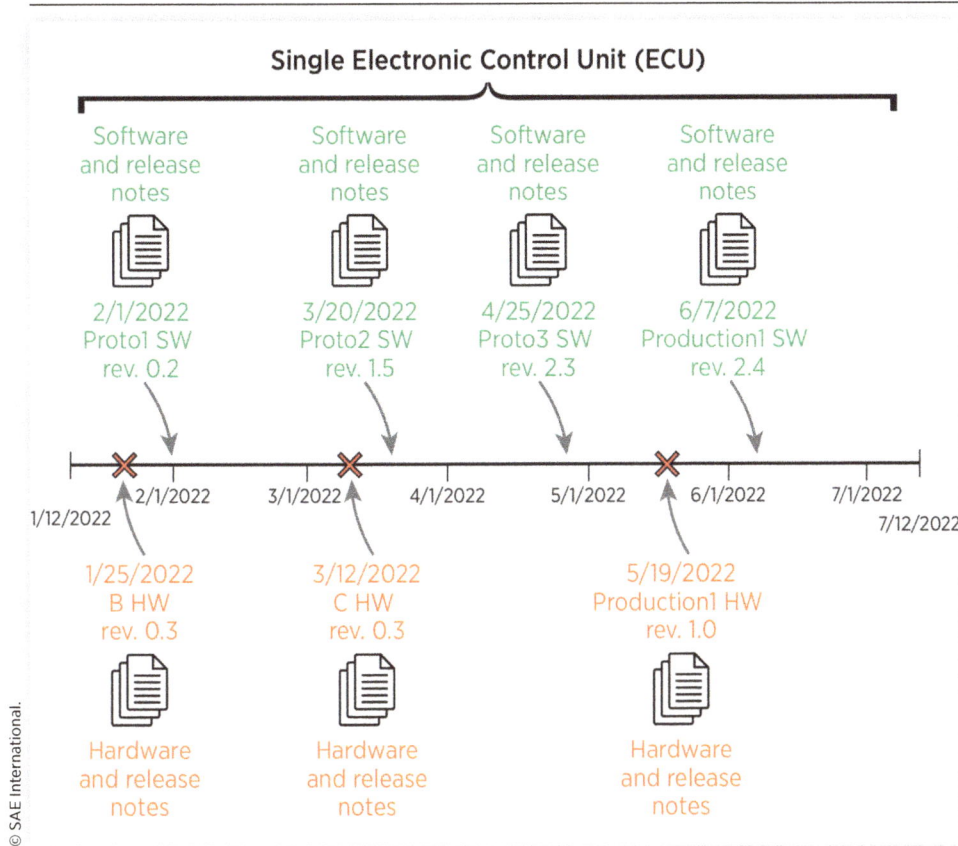

It is important to ensure that your investment and tooling readiness are up to date before investing in new tools or equipment. This means that you need to know how much money you have, what kind of technology is available, and what kind of support you will get from the manufacturer.

You can use the TRL process to determine if your organization has the right kind of equipment and support in place before moving forward with an investment or purchase.

Ability to Eliminate Low-Maturity Solutions

A low-maturity solution has been developed but has not been tested yet. This can mean that it is not fully tested, or it may have some bugs or issues that need to be addressed.

The maturity level of a solution is determined by its age. The older a solution is, the more mature it is. A solution that has been around for a long time can be considered mature and ready to be used in production. However, some solutions may not be as mature or reliable as others because they have not been tested extensively. These solutions may cause production errors or even completely fail while using your team.

This is where TRL comes into play. It allows you to stop using these low-maturity solutions by increasing the number of tests you run on them before implementing them into your workflow. If an error occurs during testing, it will be easier to fix and prevent future errors from happening again when working with the same code base.

If you are in a situation where you have a low-maturity solution, you should use TRL to help eliminate it. The reason for using TRL is to get rid of the low-maturity solution and focus on creating something new instead!

Technology Readiness Levels

In the innovation ecosystem, and particularly in innovation financing, it is important to be able to quickly and universally interpret the state of technological maturity of a project. The TRL is used as a measure (Figure 5.13). A product TRL runs from 1 to 9, with 1 equating to the early stages of conceptualization and 9 to full market deployment.

FIGURE 5.13   Technology readiness levels.

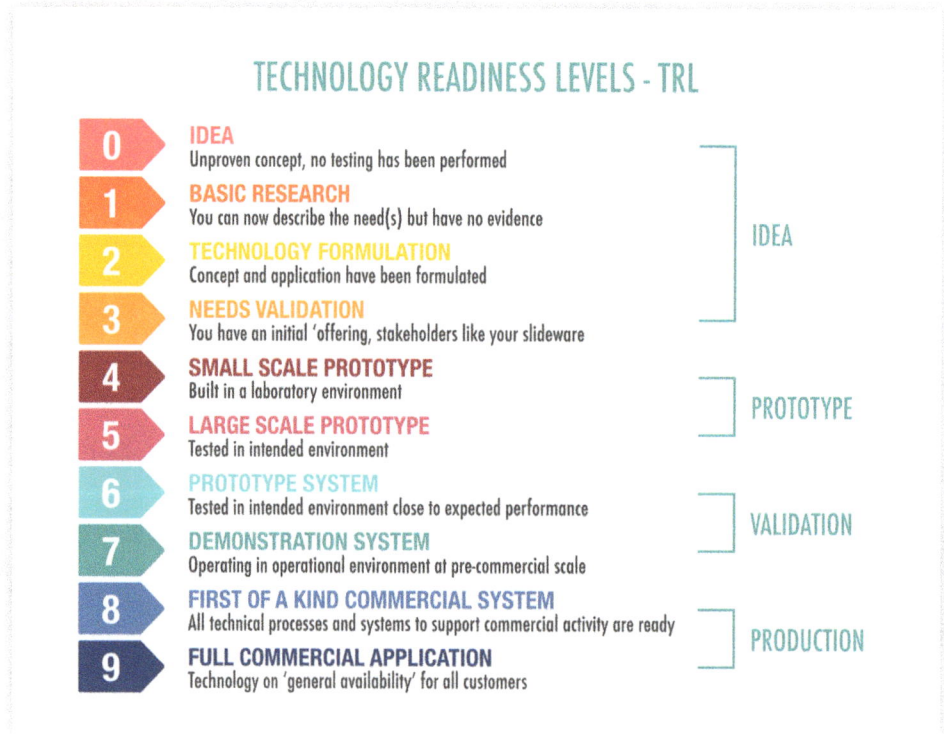

TECHNOLOGY READINESS LEVELS - TRL

0 IDEA
Unproven concept, no testing has been performed

1 BASIC RESEARCH
You can now describe the need(s) but have no evidence

2 TECHNOLOGY FORMULATION
Concept and application have been formulated

3 NEEDS VALIDATION
You have an initial 'offering, stakeholders like your slideware

IDEA

4 SMALL SCALE PROTOTYPE
Built in a laboratory environment

5 LARGE SCALE PROTOTYPE
Tested in intended environment

PROTOTYPE

6 PROTOTYPE SYSTEM
Tested in intended environment close to expected performance

7 DEMONSTRATION SYSTEM
Operating in operational environment at pre-commercial scale

VALIDATION

8 FIRST OF A KIND COMMERCIAL SYSTEM
All technical processes and systems to support commercial activity are ready

9 FULL COMMERCIAL APPLICATION
Technology on 'general availability' for all customers

PRODUCTION

Dimitrios Karamitros/Shutterstock.com.

TRL is a systematic metric/measuring system that enables assessments of the maturity of a specific technology and the consistent comparison of maturity across other types of technology.

It has evolved and is now used outside of the National Aeronautics and Space Administration (NASA). The Department of Defense (DOD) adopted TRL in 2002, and its versions are used by the Department of Energy and the European Space Agency. **A simplified scale is sometimes used for context and explanation:**

- **Conjecture:** This is TRL 1. At this level, hypotheses are formed, and an initial scientific investigation is initiated.

- **Speculation:** TRL 2 is reached when the basic principles have been well studied and practical applications can be attempted. These applications are speculative because there is little experimental proof of concept for the technology.

- **Science:** TRL 3 is when active research and design are done. Additional studies are carried out to determine if the technology is viable.

- **Technology:** When proof-of-concept models are built and tested, we can move on to TRL 4 where multicomponent parts are tested against each other. This process is called TRL 5 when dashboard technology and simulations in more realistic environments begin. TRL 6 is reached when a fully functional prototype/model is built. For NASA, TRL 7 requires that the prototype/model be tested in the space or airborne environment (whichever is appropriate for the end use of the technology).

- **Application:** TRL 8 is reached when the technology has been tested and qualified to fly. It is ready to be integrated into an existing technological system and to be implemented. TRL 9 is achieved when the technology has been flight proven during a successful mission.

Basic Understanding of Physics

Like all basic sciences, physics starts from experimental observations and quantitative measurements. Its main objective is the development of laws that govern natural phenomena to develop theories that can predict the results of future experiments.

Physics: The Most Fundamental of Sciences Physics is the most fundamental field in the sciences and is the foundation of other sciences, such as astronomy, chemistry, and geology. Physicists want to understand how nature works at all levels. This includes all aspects that extend from the size of less than the atomic nucleus to beyond the size of a galaxy. Everything is included: the elementary particles in the atom, living cells, solids, the human brain, the planets, and the universe itself. Physics aims to explain the behavior of the world around us and its laws. These fundamental laws and forces are expressed in the mathematics language, the tool that allows a direct connection among theory, practice, and experimentation.

The Beauty of Physics The beauty of physics is in the simplicity of its fundamental theories and in the way that some concepts, equations, and assumptions can alter and expand our view of the world around us. Physics is a science that is founded on experimentation and quantitative measurement.

Discrepancy between Experiment and Theory When a discrepancy between experiment and theory arises, new theories and experiments arise to eliminate the discrepancy. Sometimes a theory is satisfactory under a limited number of conditions, and it is then necessary to formulate a new, more general theory without limitations. This is the case with the so-called classical physics and modern physics. Classical physics developed before 1900 includes theories, laws, and concepts in mechanics, thermodynamics, and electromagnetism. These theories and laws were limited in many respects. The development of technology in this new 20th century led physicists to develop more general theories, ushering in a new era in physics called modern physics. A classic example is Newton's laws of motion.

Newton's Laws of Motion Isaac Newton's laws of motion describe the motion of objects at relatively low speeds. However, they do not describe the motion of objects when they reach speeds close to the speed of light. From this discrepancy arises the theory of relativity formulated by Albert Einstein.

Areas of Knowledge in Physics   Physics includes five areas of knowledge:

- **Mechanics** is concerned with the movement of objects.
- **Thermodynamics** relates to heat, temperature, and the behavior of large numbers of particles.
- **Electromagnetism** includes the theories of electricity, magnetism, and electromagnetic fields.
- **Relativity** is a theory of the movement of particles at speeds close to the speed of light.
- **Quantum mechanics** is related to the behavior of particles at microscopic and macroscopic levels.

Potential Application of Physical Phenomena

Physical phenomena or changes are changes in the state of matter that occur without altering its chemical composition since they do not involve chemical reactions. They are distinct from chemical phenomena in the latter. They are mostly reversible.

Physical phenomena include the collection of forces that typically impact matter and its change of aggregation state: liquid, solid, gaseous, or plasma. They can also be related to material mixes as long as they are heterogeneous mixtures in which the solvent and solute do not form any permanent molecular bond.

Physical phenomena may, in principle, be seen with the naked eye since the state of matter changes macroscopically. This is especially true for physically reversible alterations.

However, the amount of matter is not affected in this sort of phenomenon, implying that the change does not indicate a significant alteration of it, nor the production or destruction of it, but merely the passage from one state to another, or from one structure to another.

Experimental Proof of Concept (Demonstration)

Physical phenomena, such as heat and electricity, can be used to demonstrate the application of new technology. For example, in a case study, we will show how physical phenomena can be used to demonstrate the feasibility of new technology. To do this, we will use a system consisting of an electric circuit (a path for current) and some electrical components that can generate energy. This system will then demonstrate the feasibility of our proposed idea by generating energy without using any batteries or other devices that would consume excessive amounts of energy.

Physical phenomena are an important part of our everyday lives. We use them for cooking, cleaning, transporting ourselves and our goods, and even getting the work done. Many individuals, however, are unaware that humans may influence these physical phenomena.

With modern technology, it is possible to create physical phenomena that can be used to accomplish tasks or accomplish tasks differently than they were originally intended to be used. These phenomena are called "physical mechanisms."

Laboratory Verification

Physical phenomena are important to the application of Laboratory Verification (Figure 5.14). To successfully verify physical phenomena, it is necessary to have a thorough understanding of the system being verified and its relationship with other systems. This requires a thorough knowledge of the system itself, including its components, subsystems, their relationships, and how they are interconnected.

FIGURE 5.14 Laboratory and product certification to certified product.

Certified   Certified product   Lab tested   Tested product

LiubouS/Shutterstock.com.

After gathering this information, it must be organized so that it can be understood by different people working on the same project. This requires a clear understanding of how each person's responsibilities relate to each other's responsibilities and how their tasks contribute to the overall goal or objective of the project. This process must also be communicated clearly among team members so that everyone understands what needs to be accomplished for verification to succeed.

Breadboard or Prototype Verification in Environment

In environmental engineering, there is a need to verify the performance of new systems and devices. In some cases, this requires an early prototype or "proof-of-concept" system to test the functionality and safety of new technologies.

To test a technology in a real-world environment, it is important to understand how it will behave when exposed to various conditions. To do this, engineers use physical phenomena in their designs. For instance, if you want to ensure that your prototype does not topple over when relocated from one area to another, then you would use gravity as one of the factors that could affect its stability. If your device relies on gravity to operate properly as part of its design, it must also be able to withstand other types of forces, such as wind or water pressure.

This approach is called "hysteresis testing" because it involves applying different types of forces (or stressors) over time to determine whether your device will fail under those conditions.

Technology Demonstration in Relevant Environment

The physical phenomena now being studied and implemented in technology can be applied to the real world. For example, suppose we can develop a way for a robot to interact with human beings. In that case, we could use this technology to create robots that will be able to interact with their human counterparts without any problems at all.

This is just one example of how physical phenomena could be applied in the future by using technology demonstration in relevant environments.

System Validated in Operational Environment

Using physical phenomena is an effective way to validate system performance because it allows the user to observe how a system functions. This method can be used in many different fields, including medicine and engineering.

We have seen that many different types of physical phenomena can be applied to a system. This is because the system itself is affected by the physical phenomenon, and the effect on the system can be used as a metric for evaluating whether or not that particular phenomenon is suitable for use in an operational environment.

System Complete and Qualified

The potential applications of physical phenomena through a complete and qualified system can be used in many different applications (Figure 5.15). One such application is to create a system that can detect when the system is running out of resources. This can be useful for detecting when something has gone wrong or when there is a potential threat to the machine integrity.

FIGURE 5.15   System must be intact to qualify.

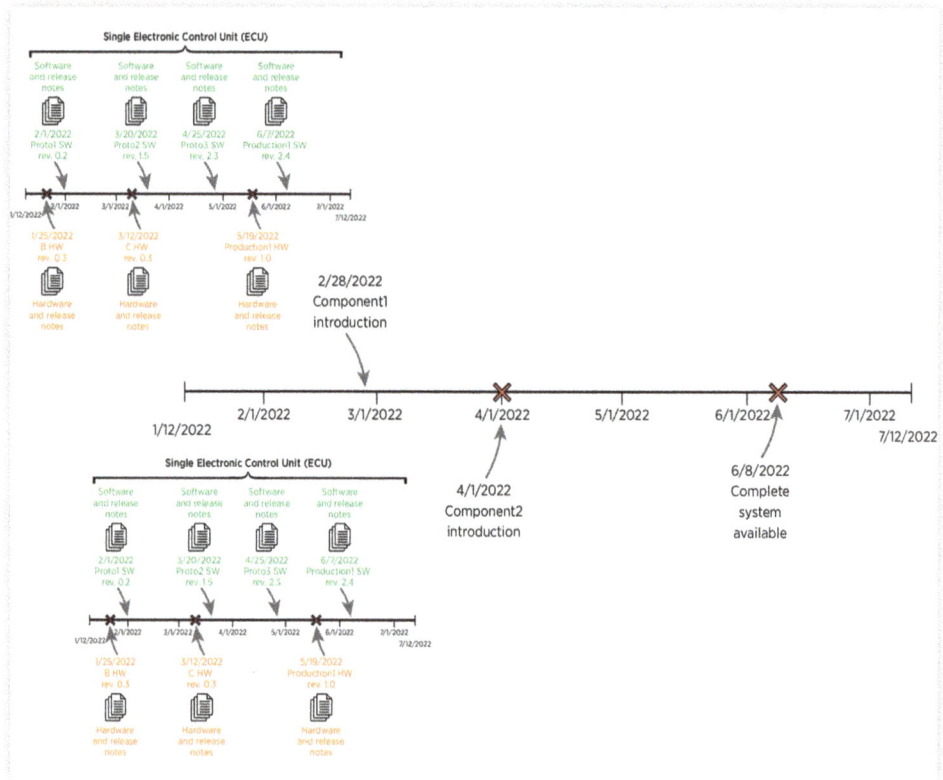

The basic idea behind this approach is that we can use the natural tendency of physical systems to seek equilibrium to guide how they will respond to environmental changes. In other words, if there is an imbalance in nature, it will seek equilibrium by correcting it with other forces or by rebalancing itself. This makes it easier for us to predict what will happen next and make informed decisions about how best to deal with situations like this.

System Validated in Operating Environment (Live Fire)

The potential application of physical phenomena by a system validated in an operating environment (live fire) is the ability to assess the performance of a system in a real-world environment. The ability to assess the performance of a system can be used to identify potential solutions or areas for improvement, as well as determine if there are any problems with the system.

There are many different methods used to assess the performance of a system. These methods include parametric testing and experimental design, and both are used when assessing systems based on physical phenomena. Parametric testing involves using statistical models to predict a system performance in various conditions based on its size and shape. In contrast, experimental design involves using controlled experiments to determine how well a system performs under different conditions.

For these methods to be effective, they must be able to predict exactly how well systems will perform based on their size and shape and determine how well they will perform under different conditions (i.e., temperature). This can only happen if data are available from previous tests conducted on similar situations at other locations or times.

TRL and Verification

The TRLs are a set of documents describing the technology readiness to be used in production. They are developed by the Defense Advanced Research Projects Agency (DARPA), a part of the DOD. The TRLs are updated annually and divided into three categories: basic research, technology development, and operational use. In this section, we will discuss TRLs and verification.

Component

The TRL is a specific type of measurement that is used to describe the competence of a technology or product. The components of a TRL are

- **Competency:** This part of TRL measures the skills, knowledge, and abilities needed to use the technology. It includes competencies in using technical information, communication skills, and knowledge about the technology itself.

- **Capability:** This part of TRL identifies the resources and support a company needs to use its technology effectively. The capability assessment looks at whether there are enough skilled people and whether they are properly trained to use it.

- **Reliability**: Reliability refers to how well a product works under normal conditions (like when it is not being used). Reliability also evaluates how well a product can operate and perform over time and how long it will last before needing repairs or replacements.

Component Integration

The component integration of TRLs integrates new technology into the existing infrastructure. This can be done by creating an entirely new piece of technology or by upgrading existing systems. This process aims to ensure that all stakeholders have access to the same information and can use it in their own way.

This process is essential for organizations as they seek to improve their use of technology and make it easier for employees to do their jobs better. It also helps organizations reduce costs while maintaining high quality, security, and reliability levels.

Systems Integration Testing

The systems integration testing of TRLs is a system engineering task that is performed to assess the maturity of a software development process, system, or application. It allows developers to demonstrate that the design or implementation of their solution is ready for testing and integration into the operational environment (e.g., customer environment).

The test plan defines the scope and objectives of the TRL test. The test methodology identifies the steps needed to complete the test concerning each objective, including data collection and analysis methods and tools used during testing. The test results are reported in a report format and distributed among stakeholders who need access.

According to "Emerson Process Management. Plant Web Optics Analytics Overview," [5] the system engineering process is listed below:

1. **Systems Engineering Across the Acquisition Lifecycle**. Rigorous systems engineering discipline is necessary to ensure that the (DOD) meets the challenge of developing and maintaining needed warfighting capability. Systems engineering provides the integrating technical processes to define and balance system performance, cost, schedule, and risk within a family-of-systems and systems-of-systems context. Systems engineering shall be embedded in program planning and be designed to support the entire acquisition lifecycle.

2. **Systems Engineering Plan (SEP)**

   a. PMs shall prepare a SEP for each milestone review, beginning with Milestone A. At Milestone A, the SEP shall support the TDS; at Milestone B or later, the SEP shall support the Acquisition Strategy. The SEP shall describe the program's overall technical approach, including key technical risks, processes, resources, metrics, and applicable performance incentives. It shall also detail the timing, conduct, and success criteria of technical reviews.

   b. The DUSD(A&T) shall be the SEP approval authority for programs that will be reviewed by the DAB/ITAB. DOD Components shall submit the SEPs to the Director, SSE, at least 30 days before the scheduled DAB/ITAB milestone review.

3. **Systems Engineering Leadership**. Each PEO (Program Executive Officer), or equivalent, shall have a lead or chief systems engineer on his or her staff responsible to the PEO for the application of systems engineering across the PEO's portfolio of programs. The PEO lead or chief systems engineer shall:

   a. Review assigned programs' SEPs and oversee their implementation.

   b. Assess the performance of subordinate lead or chief systems engineers assigned to individual programs in conjunction with the PEO and PM.

4. **Technical Reviews**. Technical reviews of program progress shall be event-driven and conducted when the system under development meets the review entrance criteria as documented in the SEP. They shall include participation by subject

matter experts who are independent of the program (i.e., peer review), unless specifically waived by the SEP approval authority as documented in the SEP.

5. **Configuration Management**. The PM shall use a configuration management approach to establish and control product attributes and the technical baseline across the total system lifecycle. This approach shall identify, document, audit, and control the functional and physical characteristics of the system design; track any changes; provide an audit trail of program design decisions and design modifications; and be integrated with the SEP and technical planning. At completion of the system level Critical Design Review, the PM shall assume control of the initial product baseline for all Class 1 configuration changes.

Live Fire (Scenario Based)

TRLs are a way to measure the maturity of the technology (Figure 5.16). The TRLs are defined by the maturity level of the innovation and the degree to which it will be used in the future. For example, an innovation that has reached Level 10 would be considered mature. In contrast, an innovation at Level 0 would not be considered mature because there is no potential for its use in the future.

FIGURE 5.16   System testing also known as live fire exercises qualify the product.

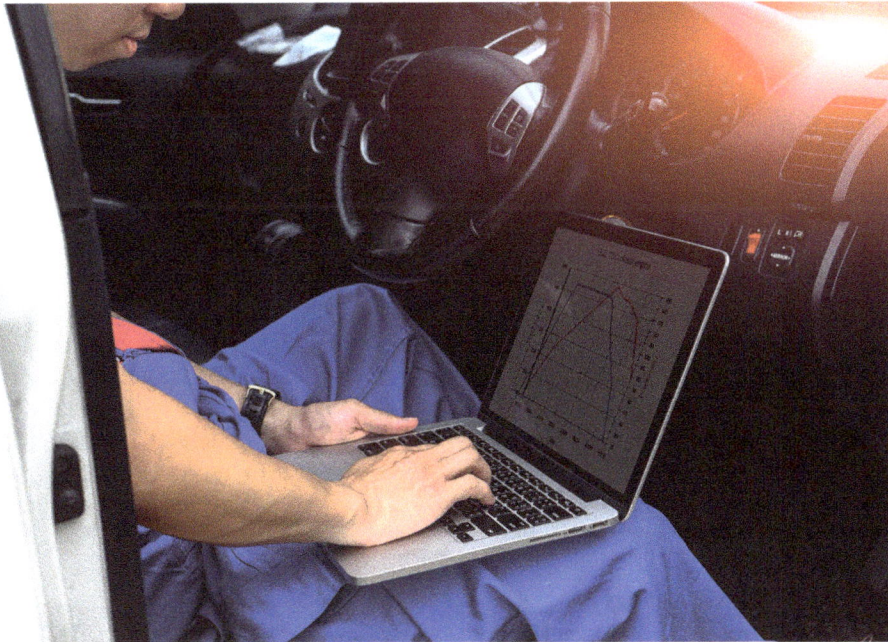

ichefboy/Shutterstock.com.

The technologies at each TRL are categorized into three categories: "ready for use," "ready for development," and "not ready for use." The first category consists of technologies that have passed all relevant tests and are ready for use. The second category comprises technologies that have passed only some relevant tests and are ready for development.

The third category consists of technologies that have not passed any relevant tests and are not ready for use.

Reasons for MRL

MRL is a system of measurement used to evaluate the readiness of a company's manufacturing facility to perform its operations at maximum efficiency (Figure 5.17).

FIGURE 5.17   Design development progress example.

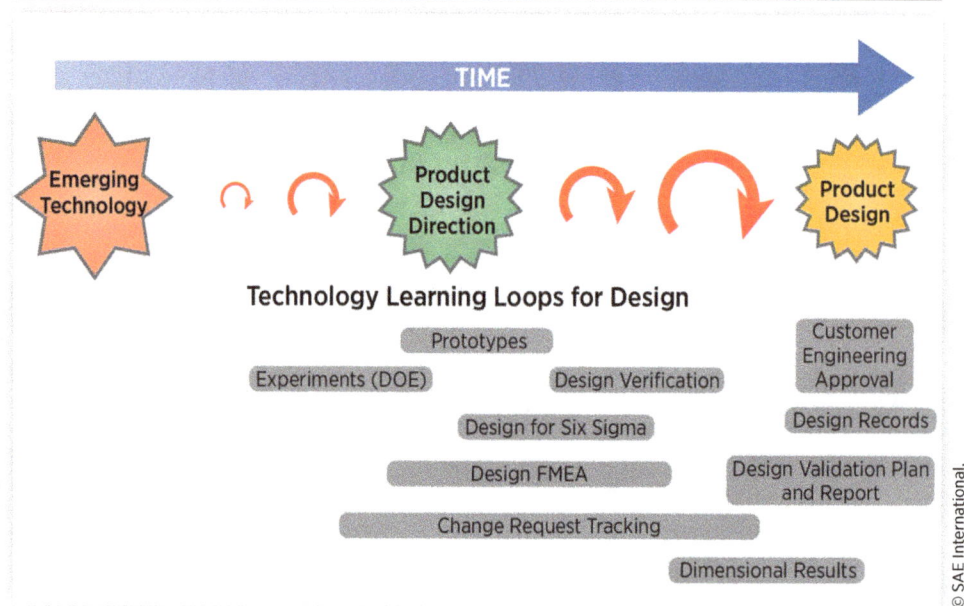

According to research file "DOD can Achieve Better Outcomes by Standardizing the Way Manufacturing Risks are Managed" (April 2010) [4], MRLs have been proposed as new criteria for improving the way DOD identifies and manages manufacturing risks and readiness. Introduced to the defense community in 2005, MRLs were developed from an extensive body of manufacturing knowledge that includes defense, industry, and academic sources. An analysis of DOD's technical reviews that assesses how programs are progressing show that MRLs address many gaps in core manufacturing-related areas, particularly during the early acquisition phases. Several Army and Air Force centers that piloted MRLs report these metrics contributed to substantial cost benefits on a variety of technologies and major defense acquisition programs. To develop and manufacture products, the commercial companies we visited use a disciplined, gated process that emphasizes manufacturing criteria early in development. The practices they employ focus on gathering sufficient knowledge about the producibility of their products to lower risks, and include stringent manufacturing readiness criteria to measure whether the product is sufficiently mature to move forward in development. These criteria are similar to DOD's proposed MRLs in that commercial companies:

- Assess producibility at each gate using clearly defined manufacturing criteria to gain knowledge about manufacturing early.
- Demonstrate manufacturing processes in a production-relevant environment.
- Emphasize relationships with critical suppliers.

However, a key difference is that commercial companies, prior to starting production, require their manufacturing processes to be in control that is, critical processes are repeatable, sustainable, and consistently producing parts within the quality standards. DOD's proposed MRL criteria do not require that processes be in control until later. Acceptance of MRLs has grown among some industry and DOD components. Yet, DOD has been slow to adopt a policy that would require MRLs across DOD. Concerns raised by the military services have centered on when and how the MRL assessments would be used. While a joint DOD and industry group has sought to address concerns and disseminate information on benefits, a consensus has not been reached. If adopted, DOD will need to address gaps in workforce knowledge, given the decrease in the number of staff in the production and manufacturing career fields.

Poor Quality

Quality is an important part of manufacturing, and poor quality can be a reason for the readiness level of an MRL. The MRL is a tool manufacturers use to determine if they are ready to produce finished goods. It considers factors like the amount of money needed to start production, how many people are required for production, and whether or not there is a market for the products being made.

Inability to Meet Production Volume

The Inability to Meet Production Volume is a reason for MRL. The MRL measures how far along your organization is in terms of readiness for production. It represents the level of commitment that a company has to be able to produce its product consistently. The MRL is determined by considering several factors such as production capacity and supply chain availability.

Factory Layout Not Aligned with Process Improvement

The reason for MRL is that the factory layout is not aligned with process improvement. This means that the factory has inefficient processes and poor quality, leading to lower production efficiency and high costs. The MRL should be used to identify and correct any problems in the current manufacturing operation.

Inconsistency in Manufacturing Processes

Manufacturers should have an MRL because it is important to ensure consistency in the manufacturing process. Inconsistency in manufacturing can lead to quality issues, ultimately affecting the company's reputation and bottom line.

The reason for this standardization was that many manufacturers used different methods or materials in their products, leading to inconsistency between some brands and models. This meant that consumers might not know what they were buying because they could not always trust that it would be safe or effective.

Operator Ergonomic Issues

The MRL was created in response to operator ergonomic issues that were causing injuries and deaths. Manufacturers use the MRL to identify the areas of their facility where operators are most likely to be injured and then design a training program for those areas.

Workforce Competency

The MRL is a measure of a company's ability to produce goods efficiently, which can be used to evaluate the readiness of manufacturing facilities and identify problems that need to be addressed.

It is no longer enough for a company to produce goods at a reasonable cost or even at a reasonable profit. Now you need to be able to do it in a way that will provide the best value for your company and its customers.

Manufacturing Dimensions

The MRL dimensions are a set of dimensions used by manufacturers to demonstrate their readiness for production. Government agencies also use these dimensions to measure MRLs, which can then determine if the country is ready for mass production. **The MRL dimensions include:**

Technology Deployment and Industrial Index

Technology deployment and industrial index is one of the MRL dimensions. It indicates how well a country's manufacturing enterprises can absorb, adopt, and adapt to technological change.

The Technology Deployment and Industrial Index measure the degree to which technology is used in production processes of manufacturing companies and the extent to which these companies are likely to adopt new technologies.

Line Design

Line design involves the process of determining the shape and size of a product. This process is commonly referred to as "design." The line design dimension sets requirements for the minimum number of fixtures, devices, or tools (or any combination thereof) present at a designated workstation during a specific period.

Available Resources

Available resources are one of the MRL dimensions. In this dimension, the manufacturer determines how many resources they have, such as raw materials, personnel, and equipment. For example, suppose a manufacturer only produces a single product and their employees are highly trained in that area. In that case, they will have more available resources than someone who lacks this knowledge.

Material

Material is one of the MRL dimensions. The Material Readiness Level is a prerequisite to a company's ability to produce products or services and is one of five readiness levels defined by the American Society for Quality.

The Material Readiness Level represents the readiness of a company's resources to perform the planned production tasks. It is determined by measuring the capability of an organization to produce materials and processes per customer requirements and standards.

According to "Manufacturing Readiness Level Deskbook" [6], successful manufacturing has many dimensions. MRL threads have been defined to organize these dimensions into nine manufacturing risk areas. The threads are as follows:

- Technology and the Industrial Base: Requires an analysis of the capability of the national technology and industrial base to support the design, development, production, operation, uninterrupted maintenance support of the system and eventual disposal (environmental impacts).

- Design: Requires an understanding of the maturity and stability of the evolving system design and any related impact on manufacturing readiness.

- Cost and Funding: Requires an analysis of the adequacy of funding to achieve target manufacturing maturity levels. Examines the risk associated with reaching manufacturing cost targets.

- Materials: Requires an analysis of the risks associated with materials (including basic/raw materials, components, semi-finished parts, and subassemblies).

- Process Capability and Control: Requires an analysis of the risks that the manufacturing processes are able to reflect the design intent (repeatability and affordability) of key characteristics.

- Quality Management: Requires an analysis of the risks and management efforts to control quality, and foster continuous improvement.

Manufacturing Management: Requires an analysis of the orchestration of all elements needed to translate the design into an integrated and fielded system (meeting Program goals for affordability and availability). Many of the MRL threads have been decomposed into sub-threads. This enables a more detailed understanding of manufacturing readiness and risk, thereby ensuring continuity in maturing manufacturing from one level to the next. For example:

- Materials includes maturity, availability, supply chain management, and special handling (i.e., government furnished property, shelf life, security, hazardous materials, storage environment, etc.).

- Process Capability and Control includes modeling and simulation (product and process), manufacturing process maturity, and process yields and rates.

- Quality Management includes supplier quality.

- Manufacturing Management includes manufacturing planning and scheduling, materials planning, and tooling/special test and inspection equipment.

Manufacturing Process Capability and Control

Manufacturing Process Capability and Control is one of the MRL dimensions. The Manufacturing Process Capability and Control dimension focuses on a company's ability to manufacture products cost-effectively. It measures a company's ability to produce products following customer requirements using efficient manufacturing processes, materials, and equipment.

Quality Management, Including Supplier Quality

The dimension is concerned with a manufacturer's ability to produce consistent products that meet quality specifications. A product quality is directly related to its ability to perform as intended by its design. For a product or service to be successful, it must be able to meet the needs of an end user.

Manufacturing Workforce Competencies (Engineering and Production)

The manufacturing workforce competencies dimension describes the knowledge and skills required to perform a specific role in the manufacturing process effectively. In this example, engineering competencies are required to operate machinery and perform analysis and design.

Facilities

Facilities are a key component of any manufacturing operation, and they can make or break a company's ability to turn out products on time and within budget. Facilities include everything from buildings and equipment to trucks and forklifts.

Facilities are a key component of a manufacturing facility's ability to achieve its operational requirements. In other words, people use facilities to perform the functions necessary for fulfilling the MRL dimensions of process capability (process efficiency) and equipment condition (equipment functionality).

Manufacturing Readiness Level

The MRL is a tool used by the US government to measure the readiness of a plant to manufacture on an assembly line. It is reported by each manufacturer to the federal government and helps determine how much funding from the government is available to be given to a business for them to become more competitive and efficient.

The MRL measures how prepared a company is for production and how successful it will be at meeting its goals based on what it needs from its facility for it to be productive. The level of MRL can range from 0 to 99, with higher numbers indicating greater readiness and efficiency than lower ones.

As companies start with low levels of MRLs, they may have difficulty receiving funding from the government because there are not enough businesses that meet those levels yet; however, once they have reached higher levels, they will receive more funding so that they can continue progressing toward becoming more efficient and productive in their facilities while still having room for growth.

According to Final rule of DOD, Federal Register [7], DOD is adopting as final, without change, an interim rule amending the Defense Federal Acquisition Regulation Supplement to implement a section of National Defense Authorization Act for Fiscal Year 2011 requiring appropriate consideration of the manufacturing readiness and manufacturing-readiness processes of potential contractors and subcontractors as a part of the source selection process for major defense acquisition programs. DOD published an interim rule in the Federal Register at 76 FR 38050 on June 29, 2011, to amend Defense Federal Acquisition Regulation Supplement (DFARS) 215.304(c) by adding paragraph (iv) to state that the manufacturing readiness and manufacturing-readiness processes of potential contractors and subcontractors shall be considered as a part of the source selection process for major defense acquisition programs. No public comments were submitted in response to the interim rule.

Basic Manufacturing Implications

The basic manufacturing implications of the Fourth Industrial Revolution are that it will increase automation and AI, creating many jobs. However, it also means that there will be a decrease in jobs related to manual labor and those who perform tasks using creativity and innovation (Figure 5.18).

FIGURE 5.18 Example of manufacturing readiness activities.

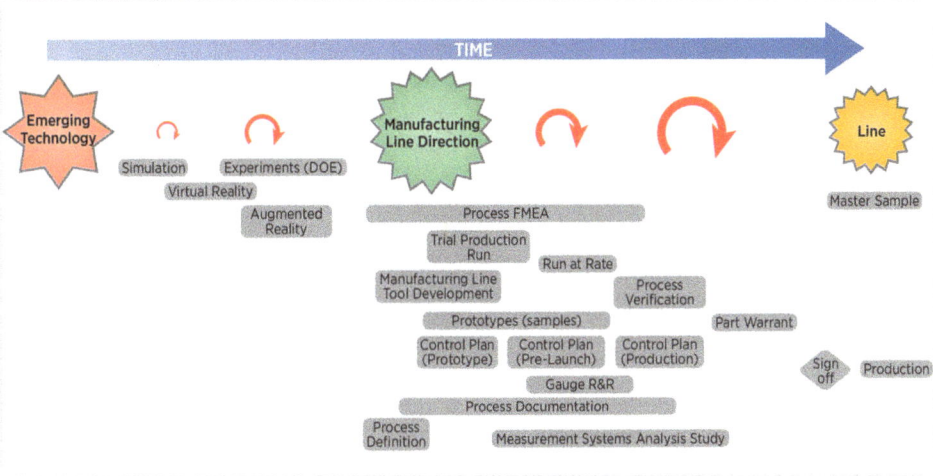

© SAE International.

Manufacturing Concepts Identified

The concept of manufacturing is based on the idea that the product being made is created by combining various raw materials. The manufacturing process involves extracting raw materials from their source, processing those materials, and then combining them into the final product.

Manufacturing Proof of Concept Developed

The manufacturing proof of concept is a method for determining whether a specific product can be developed and manufactured cost-effectively. It involves performing several tests to ensure that the product design and production process have been successfully analyzed and that all costs are considered when determining whether or not the product will be profitable.

Ability to Produce Technology in the Laboratory Environment

The ability to produce technology in the laboratory environment is a prerequisite for many jobs, and it is also an important skill for anyone who wants to develop new technology. This is because your knowledge of how things work will help you make better decisions about what you want to build next. It is also a great way to learn more about other fields within technology. For example, if you are interested in biology, you might want to look at how biotechnology works.

Ability to Produce Prototype Components in Relevant Environment

A prototype is a representation of an idea used as a basis for designing a new system. It is made to test various aspects of the new system, which can be used to verify that the design can be built and function successfully as intended.

The ability to produce a prototype component in an appropriate environment is extremely important because it allows designers to test their ideas in real life. This means they will have more confidence when presenting their designs to clients, leading to better outcomes for everyone involved.

Ability to Produce a System in a Representative Production Environment

The ability to produce a system in a representative production environment is the ability to create a system that will work in a real-world setting. In other words, you need to be able to create something that works independently without being interfered with by other factors or people.

This is especially important when it comes to computer programming. If you are going to make a program that runs on your computer or phone, you need to make sure that it can run in all conditions—no matter what the weather is doing outside or how much time you have before an exam. You need to write code that works with whatever hardware and software are available at the time (and possibly even later ones).

Pilot Line Capability Demonstration—Low-Rate Production

The pilot line capability demonstration is a low-rate production process that allows a company to test the feasibility of their new product and determine whether it will work in mass production. This process involves creating a small amount of product that can be used for testing purposes and is usually conducted individually.

The goal of the pilot line capability demonstration was to demonstrate that it was possible to produce aircraft parts at low rates using existing technology. The goal was not to produce an entire plane but just one part of the plane, such as a wing or tail section. With this knowledge, manufacturers could use this data to scale up their production processes for other parts.

Low-Rate Production Demonstration in Place of Production Ramp to Full Rate

The low-rate production demonstration is a way to show how the company will operate in a slow-growth environment. This can be done by using a set of products or services that are important to the company's business and then slowing down the production rates of those items. This can be done by adjusting the number of items produced per hour, increasing or decreasing the total amount of time spent on each product, or changing the number of people working on each product.

If the company would like to continue growing its business, it will need to increase its production rate and/or change some other factors to keep up with demand. If this is not possible for some reason, then the company should reduce expenses (such as wages) to continue operating at its current level despite slower growth rates.

Full Rate Production Demonstration in Place

In the Full Rate Production Demonstration, workers are given a fixed quantity of product to produce in a fixed period. They then receive additional resources as they go along, so the rate at which they can produce increases over time.

In this way, it is like a production line: workers are given a fixed amount of material and then receive more as they go along. They work at the same rate (or "rate of production") throughout the demonstration, even if they add multiple parts or remove parts from the line. The purpose of using this method is to show how much more efficient it is for workers to work at higher rates than their usual rates.

References

1. U.S. Government Accountability Office, "Best Practices: Capturing Design and Manufacturing Knowledge Early Improves Acquisition Outcomes (GAO-02-701)," July 2002.

2. Wheeler, D. and Ulsh, M., "Manufacturing Readiness Assessment for Fuel Cell Stacks and Systems for the Back-up Power and Material Handling Equipment Emerging Markets," Technical Report NREL/TP-560-45406, United States Department of Energy, National Renewable Energy Laboratory, February 2010.

3. Defense Acquisition Regulations System, "Interim Rule, 76 FR 38050," Federal Register, June 29, 2011.

4. U.S. Government Accountability Office, "Best Practices: DOD can Achieve Better Outcomes by Standardizing the Way Manufacturing Risks are Managed (GAO-10-439)," April 2010.

5. EMERSON, "Emerson Process Management—Plant Web Optics Analytics Overview," 2020.

6. DOD, "Manufacturing Readiness Level Deskbook," May 2, 2011.

7. Defense Acquisition Regulations System, "Final Rule, 76 FR 71645," Federal Register, November 18, 2011.

6

The Right Solution

Determine Solution

Why Wait?

Clarify Objectives and Priorities

The extent to which a specific project meets its initial goals constitutes the project execution success metric we use. This is not relegated solely to the design but also the manufacturing approach that is taken. Technical performance, product functionality and quality, unit cost, and time to market for the development effort are the main goals of a PD project. These goals are established before the commencement of the project execution, and the success of those goals is assessed at the project's conclusion. This performance metric stands out in four crucial ways that are consistent in evaluating the success of the PD project. First, it is a success indicator for the project execution phase alone as opposed to project planning success or success for both planning and execution combined. Second, it is an internal measure. Third, PD success is unquestionably measurable; nevertheless, the internal, execution-focused scope and goal of the current research are outside the realm of market-oriented and other external measures. Fourth, a well-executed project could have a high project execution success but still produce a subpar product for the market (Figure 6.1). If the product is poorly planned, for example, if the incorrect product features were chosen or poorly introduced into the market, insufficient sales promotion, or another market failure may still occur despite high operational success.

FIGURE 6.1 Delay solution selection reduces regret.

Project Execution Success The project execution success metric is the weighted average of how well the three main project goals of technical performance, product unit cost, and time to market were each accomplished. The weights are based on how important each goal was for the project. Given that many projects have various objectives, such a measure is better able to capture project-to-project subtleties [1].

Develop Context

NPD frequently crosses organizational and regional boundaries. The complexity of the problem is illustrated in the literature on managing knowledge barriers in the inter-organizational development of ITs, which is still in the early stages of research. Surprisingly, little research has been done on the effects of IT on knowledge sharing in global NPD efforts.

When NPD occurs in a global setting, knowledge must be shared across organizational, geographical, cultural, and linguistic barriers. Because of the many hurdles to information exchange, there is a great danger of failure, which necessitates using boundary spanners. This complex circumstance has recently fueled the need for digital solutions like Organizational Memory Systems, Cooperative Work Systems, and Project and Resource Management Systems.

The market has met the requirement for integrating these systems with Product Lifecycle Management (PLM) technology, which uses object storage and processes to manage product and project information. From experience, one tool that supports coverage throughout the PD activities is more advantageous than a collection of disparate systems with limited or no connectivity.

These tools help define and standardize processes and information objects, helping the development phase from design to industrialization. Product Data Management (PDM) systems, which are devoted to managing links between product components and, more generally, engineering activities, are where Product Lifecycle Management (PLM) technology originated. Notably, PLM mainly supports development tasks and product instantiations through the development process. PLM hence has little to do with irrational or vaguely defined research activities.

Although PLM technology allows users to store and modify objects like product representations and supports development planning and scheduling, the degree of flexibility attributable to such integrated technology may be lower than the literature suggests, which hinders learning and intercultural communication. We are particularly interested in examining if and how PLM may be used and how well information is shared in a system where it transcends many different sorts of borders [2].

Decision-Making and Risk

Three Main Issues Exist in Multi-criteria Decision-Making (MCDM)

- Choice: Select the most suitable option. When faced with a simple decision-making challenge (Figure 6.2) like selecting one or more of the best options, it can help to start by screening out any options that do not seem important enough to keep in mind. A method called screening is used to condense many options into a manageable number that (presumably) includes the best option. On evaluation scales, there should be certain elimination criteria or intervals.

- Ranking: Sort all options in order of greatest to worst. It is intended to establish a global evaluation model for each alternative, accounting for all factors and scales that were considered.

- Sorting: Sort every alternative into various, previously established groupings. We consider screening and rating categories in this work. A multi-criteria decision-making issue arises after the sequential procedure of:

  - Establishing decision goals.

  - Selecting and organizing options with potential dependencies.

  - Establishing and organizing standards with potential dependencies.

  - Using weights and thresholds while considering the criteria.

  - Screening out options that do not meet criteria thresholds and analyzing alternatives for each criterion.

  - Ordering the remaining options based on their individual ratings and weighted criteria.

  - Making a choice [3].

FIGURE 6.2   Design and manufacturing decisions.

EtiAmmos/Shutterstock.com.

Build Product Knowledge

Experimentation Simulation

Techniques for virtual simulation are capable of much more than only lowering costs and accelerating problem-solving. However, they also bring about significant changes in how problems are solved. In addition to the effects of price and lead time, we believe adjusting the organization is necessary to fully benefit from virtual simulation technologies.

This argument is based on earlier studies that showed the significance of organization in NPD performance. Understanding how virtual simulation technologies impact problem-solving will help with that. The fact that the NPD process is not always compatible with the cost and time requirements of physical trials is one of the main reasons for the introduction of virtual simulation tools.

From experience, engineers frequently choose to "carry over" components and systems of components from the previous model generation to the new one because of these later factors, which results in conservative design. Significant time, cost, and quality benefits are obtained via carry-over approaches.

Additionally, engineering solutions used in prototypes are frequently outdated by the time they are constructed. Thus, the outcomes of physical experiments are often not entirely applicable to engineering development. Using virtual development tools helps you get over these constraints. They not only assist in lowering experimentation costs by accelerating the testing phase, lowering the quantity of pricey physical prototypes, and reducing redesign because of their rapid obsolescence. They also enhance design quality by making reliable information available early in development.

Engineers can see otherwise considerably fewer observable events thanks to virtual tools. However, there are limits, and there are still reasons for prototype parts. Furthermore, these virtual tools are only as reliable as the verity of the model parameters. Errant parameters are missing parameters; it is near impossible to understand all environmental and parameter interactions perfectly.

To demonstrate the point, consider the number of physical prototypes that are to be constructed for crash tests is severely constrained because of economic considerations (Figure 6.3). However, it is also critical that automobile models achieve a high score in the New Car Assessment Programme (NCAP) crash test (http://www.euroncap.com). It is easier to pinpoint the cause-and-effect correlations among all the components after repeating the experiment multiple times, given that there are at least 400 components (according to our conversation with a PD engineer) that affect such a crash test. Because of this, engineers and designers of automobiles must balance costs and experimentation accuracy.

FIGURE 6.3   Prototypes help the team learn and drive the design of the product and manufacturing line.

Virtual simulation techniques have made it possible to repeat the same experiment virtually infinitely while, crucially, isolating a single parameter in each run. Therefore, they take a "laboratory-type controlled environment" approach instead of physical experimentation. This allows for testing a wide range of hypotheses on the causal link between design attributes and crash test results (at a reasonable price) [4].

Exploring Multiple Alternatives (Learning)

Because tensions are a natural part of PD activities and drive innovation, they cannot be completely eradicated.

Successful PD entails managing tensions—coping with changing circumstances to stimulate innovation efficiency. We looked at various characteristics of project management approaches, dynamics, and effects to better understand this difficulty. Product creation has the potential to be a key driver of organizational renewal and competitive advantage, yet success is frequently elusive. PD programs cause conflicts between demands from the outside world and internal capabilities, the need for spontaneity and structure, and the need for change and stability. Project teams work to increase their capacity for innovation while also learning technical skills and achieving business goals. However, effective execution is also necessary for success to maintain projects on schedule and under budget [5].

Decide Later Benefits

Businesses across a wide range of industries are working to speed up the creation of new goods from concept to consumer, increase their quality, lower cost, and make the launch process easier. Prior studies have suggested that including material suppliers in the NPD cycle can significantly impact accomplishing these objectives. This engagement can take many forms, from simple design ideas consultation to giving vendors complete design control over the systems or components they will deliver. Additionally, suppliers could be involved in various phases of creating a new product. Supply chain, material logistics, and process design benefit from early supplier involvement.

By building a "backpack" of existing and upcoming technologies and suppliers for new technologies through ongoing debates on evolving technical discoveries and standards, many companies try to manage and obtain the best technologies for application. Uncertainty about new technology and industry norms is a given in many markets. To balance the advantages of being the "first mover" with the hazards of the technology, prominent organizations keep an eye on developing new technologies and, for those that seem to have potential applications, manage their launch in new product applications.

Research-intensive collaboration frequently entails identifying and assimilating tacit knowledge that is difficult to transmit or codify. Savings on transaction costs and quick profits are typically less critical (or should be) than improvements in technical competence, tacit knowledge, or market awareness in light of rapid market change. Evaluating the business case for supplier integration is crucial for technology sharing to result in better supplier solutions. Key supplier staff and other members of the PD team must comprehend and concur on the intended benefits of the supplier integration effort in terms of cost, quality, scheduling, roles, and responsibilities [6].

Set-Based Development and Concurrent Engineering

Besides delaying the decision, exploring multiple possible alternatives is how we find the optimum solution. With set-based PD, we do not hone in on one specific solution but explore various possible solutions. These solutions might employ the same technology, but not necessarily or not by requirement. We will spend time considering and testing each of these to arrive at the best solution, and that solution may end up being a solution with elements of two or more solutions together.

We extend the topic a bit further to concurrent engineering. Developing a product includes many domains and disciplines. Therefore, the best solution is not the solution that meets the customer's expectations solely. We will fail if we have a great design idea but cannot manufacture the product. Concurrent engineering requires input and coordination of all these domains.

What Is It and Why?

There has been a lot of interest in the PD process known as set-based concurrent engineering (Figure 6.4), or SBCE. Some writers say SBCE is four times as efficient as conventional phase-gate procedures. However, despite the substantial volume of related research, the fundamental study at Toyota Motor Corporation, where the approach was created and discovered, remains the only published use of SBCE.

FIGURE 6.4 Set-based PD.

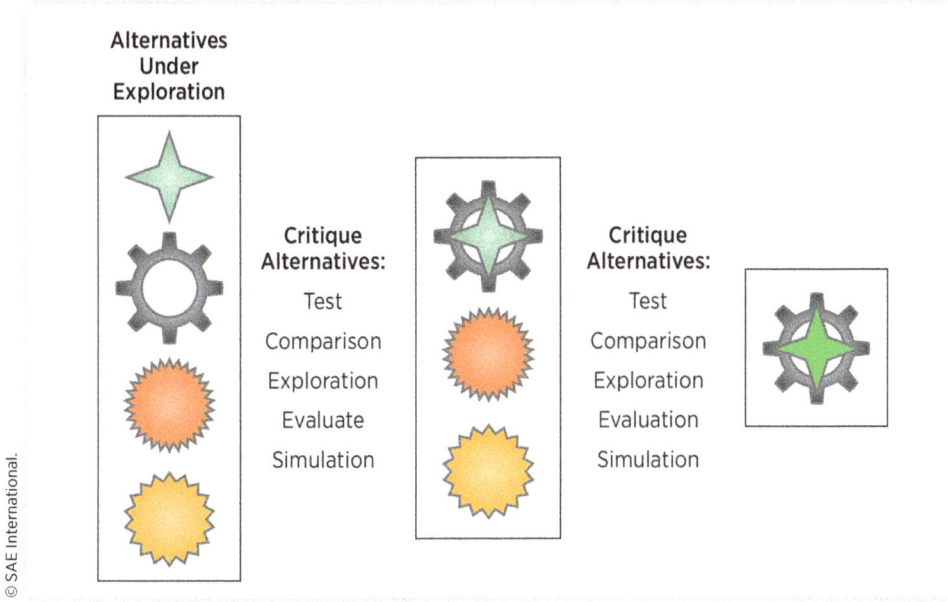

© SAE International.

The "Lean Product Development System" considers it to be a highly integrated component (Figure 6.5). The lean development method employs several techniques for managing employees, projects, and decision-making. In lean development, "set-based" refers to working with several solutions at once, methodically examining trade-offs between various choices, and utilizing visual knowledge [7].

FIGURE 6.5 Concurrent engineering includes all the product entities.

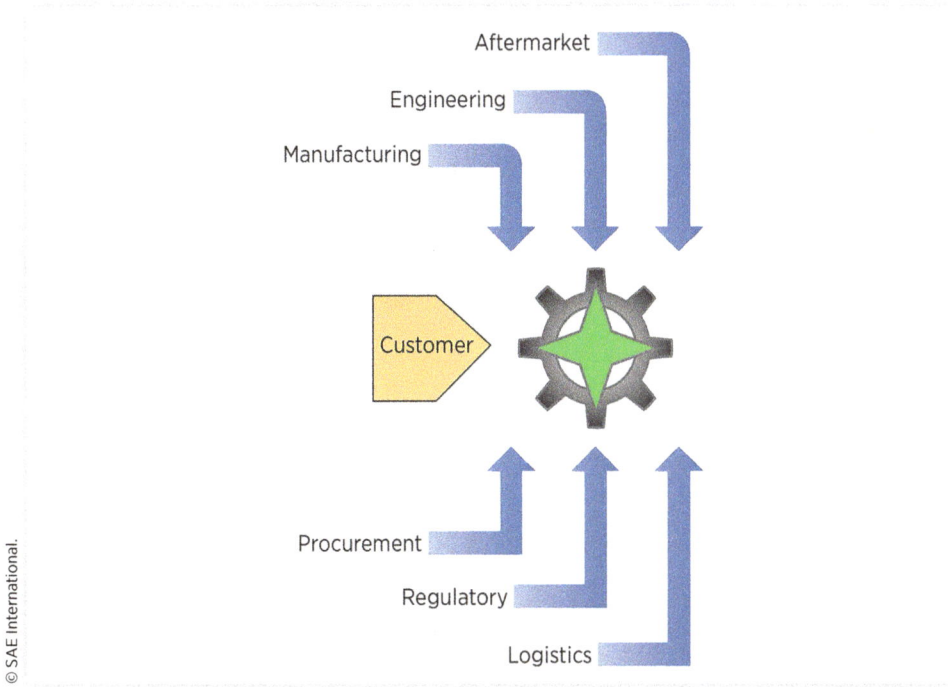

© SAE International.

Consider Multiple Versions

Component- and component-based system development differs significantly from "traditional" software system development in several important respects. The primary distinction is the division of the system development process from the component development process. This fact significantly impacts the development process. The component-based approach in software engineering is still in its infancy; therefore, the development of technologies has received most of the attention, leaving process modeling as uncharted territory. The fundamental traits of the component-based approach and how they affect the lifecycle models and the development process are examined. The general lifetime of component-based systems and the lifecycle of components are described, and the various types of development processes—product line development, commercial off-the-shelf–based development, and architecture-driven component development—are covered in detail.

The lifespan of software goods differs slightly from that of physical items since production activities are often much smaller than other activities and can, thus, be overlooked as a separate phase. Additionally, software is frequently created and delivered in several versions since it is simple to change (even though a change may have negative effects and involve a lot of work). Concurrent operation and development are made possible as the software lifecycle is performed in a slightly different way than the model for a product lifecycle.

Original development refers to the concept phase, which also includes the initial design and development. Since it is thought that the production phase is a component of the development phase, it is not included. The utilization phase, which includes continued development, is made up of several cycles for evolution and maintenance. The phase-out and close-down phases are the final divisions of the retirement period. The first operational version of the product is created entirely from scratch during the early development phase to meet initial needs. The product quality and functionality are iteratively improved during the evolution phase. New iterations of the product are supplied to consumers regularly. Only minor product flaws are fixed during the servicing phase.

During the phase-out period, the product is still in use but is no longer serviced. The product is finally taken off the market during the close-down phase and either replaced by another product or disposed of. In the early stages of development as well as during each cycle of evolution, development organizations frequently carry out the same tasks. The same series of steps will be repeated to create a new version of an existing software product, usually with a different emphasis. All these tasks taken collectively define a software development lifecycle. Not every model of the software development lifecycle is appropriate for every kind of software system.

Large systems with many stakeholders and long development cycles typically favor employing sequential models. Systems that leverage new technology are more compact, prioritize time to market, and frequently investigate evolutionary models. Consider evolutionary models frequently. These models are more adaptable than sequential models and can get results considerably faster [8].

Set-Based Concurrent Engineering

SBCE is sometimes viewed to significantly enhance the processes involved in product design. Despite its widespread use in literature, there have only been a few reported uses so far. Recent research adopted a case study approach that adds new information by describing implementations of SBCE deployments in four PD businesses. To determine whether the concepts of SBCE may enhance the efficacy and efficiency of the development process, the case study demonstrates that, with the right assistance, set-based projects can be pushed within an already-existing organization. The participants contend that a set-based strategy improves development performance, particularly in terms of innovation, cost, and performance of the final product. The gains came at the expense of a little bit more expensive development and a longer lead time. The companies involved intend to adopt SBCE in future projects when suitable because the good effects outweigh the negative ones.

The organizational implications of SBCE are significant, necessitating a change in most procedures and practices. The "Lean Product Development System" considers it to be a highly integrated component. The lean development method employs several techniques for managing employees, projects, and decision-making. In the context of lean development, "set-based" refers to working with several solutions at once, methodically examining trade-offs between various choices, and utilizing visual knowledge [7].

Competing Design and Manufacturing Alternatives

An NPD process is a set of actions or stages that a business uses to create, design, and market a new product (Figure 6.6). Many of these actions and tasks involve organizing and thought rather than physical labor. For the following purposes: quality control, coordination, planning, management, and improvement, a clearly defined development process is helpful. The development process can be conceptualized as the initial generation of a broad range of different product conceptions, the subsequent narrowing of alternatives, and the increasing specification of a product until the production system can reliably and repeatedly produce the product.

Perhaps more than any other phase of the development process, the concept development phase stretches into what is known as the "fuzzy front-end process or project planning" and necessitates collaboration among functions. The activities that make up the concept development process are as follows: determining the needs of the target market, establishing target specifications, concept generation, concept selection, concept testing, setting final specifications, project planning, economic analysis, benchmarking of similar products, modeling, and prototyping.

The PD team identifies a set of consumer needs early in the development process. The team then creates alternative solution concepts in response to these needs by utilizing several methodologies (such as QFD). The requirements of the customers are determined during the initial design stage. These client specs are used to create a list of product specifications. The specifications, which are provided in a solution-neutral format, list the features the product must offer. The following design stage, known as concept design, entails creating a set of subsystems that are compatible with one another. During concept design, several alternative subsystems are produced to execute each subset of the stated functions, and when taken as a whole, the full set can perform all the necessary functions. After these many ideas have been described, the subsystem configuration that provides the maximum performance at the lowest cost is chosen.

FIGURE 6.6   Systems thinking for both design and manufacturing while we explore alternatives.

The crucial decision in the design process is frequently concept selection. The original concepts determine the path of the design embodiment stage; thus, it is crucial to choose the greatest ones. The design process will diverge toward a detailed solution once this step has been completed. Therefore, choosing a concept is an essential step in the design process. One of the key components in the effort to boost design productivity is the capacity to quickly assess design concepts as they develop during the design process.

Given the necessity for businesses to generate ever-more innovative products in a market that is becoming increasingly competitive, designers must consequently consider a greater number of design possibilities. This is especially true in the early conceptual stages of the design process when it is essential to generate many design possibilities to guarantee that the best concept is found.

Designers face a significant problem because of the lack of industry-wide rigor and discipline needed to preserve objectivity in examining many design concepts. To help with overcoming this obstacle, design tools and aids must be made available. It is also obvious that, for any design tool or aid to be accepted in the industry, it is necessary to demonstrate the validity, dependability, and robustness of the theoretical models that underlie them.

Evaluation is the process of comparing and choosing from a variety of competing design ideas. It is obvious that human evaluators will need more support to maintain objectivity throughout a dynamic design process as the number of possibilities to evaluate grows and the time available shrinks. This is especially true during the conceptual stage of the process, when it is essential to generate, choose, and pursue the best concept for the entire design process to be successful [9].

Set-Based Integrated Product and Manufacturing Systems

Manufacturing organizations must alter their conceptual product redesign solutions in response to more stringent specific client needs. The usage of pre-made components is constrained in this circumstance, and the manufacturing system needs to be physically altered. Platforms must be designed in these circumstances with more flexibility to allow for long-term evolution. The conceptual considerations for products and production systems must be included in the respective platform models. The literature recommends functional modeling to capture these factors, but the modeling only applies to the manufacturing or product domains. Its connection to industrial processes is also not explained. Functional modeling does not, therefore, fully realize its potential to support the integrated development of goods and industrial systems.

Products that are manufactured as well as the production systems that make them are multi-technological systems made up of various hardware subsystems (e.g., mechanical, hydraulic, and electrical hardware) and software subsystems. During the various stages of their lifecycles, these systems and their subsystems engage in interactions with one another and the environment in what are known as "lifespan meetings." During and after the conceptual design phases, it is especially important to comprehend and control the interactions between the product and the manufacturing system because they determine how the product and manufacturing system interact with one another. A product change can necessitate the use of new production tools, or it might call for product modification to enable the use of a more effective manufacturing process.

Lastly, many approaches and parts can be used to represent the range of products and manufacturing systems. Upon consideration, this creates the possibility for platform thinking in the product domain to be combined with mindsets similar to those in the manufacturing domain, allowing for the integrated development of configurable product and manufacturing system platforms [10].

Right Sizing—Right Timing

High-technology companies need to take advantage of opportunities at the right time. Although the entrepreneur is the first to find the opportunity, this window of opportunity typically only exists for a brief period. As other businesses begin to capitalize on the possibilities of the new finding, opportunities could become less attractive. Alternately, rivals might support divergent technical pathways that result in breakthroughs that can serve as replacements. If the initial chance is not taken advantage of immediately, more current technology developments can also take place. Consequently, having early access to knowledge or resources that enable swift exploitation might mean the difference between finishing first and failing completely in high-technology businesses where windows of opportunity pass quickly [11].

What and Why Iteration

Companies confront multiple risks when developing new products, making the risk framework essential to PDP design. Many authors emphasize the importance of creating effective risk management plans as part of the PD process. Many studies classify development risks as being largely technical, commercial, budgetary, or schedule related. Technical risk results from the unknown ability of a new product to meet its own functional and design requirements.

Iterations, Process, Product

If the design requirements do not satisfy client expectations, a technically excellent product may not succeed in the market. Iterations, which are regulated, feedback-based redesigns, are a part of how PDPs manage risk (Figure 6.7). Small iterations might involve only slight adjustments to components, but large iterations might involve design modifications because of marketing feedback. Reviews, which serve as checkpoints or gates between development stages to ensure appropriateness, are another way that PDPs manage risk. While flexible reviews let more concurrent work, strict reviews forbid further design until preliminary work is completed [12].

FIGURE 6.7   Iterations have never been only an agile approach and address product and process.

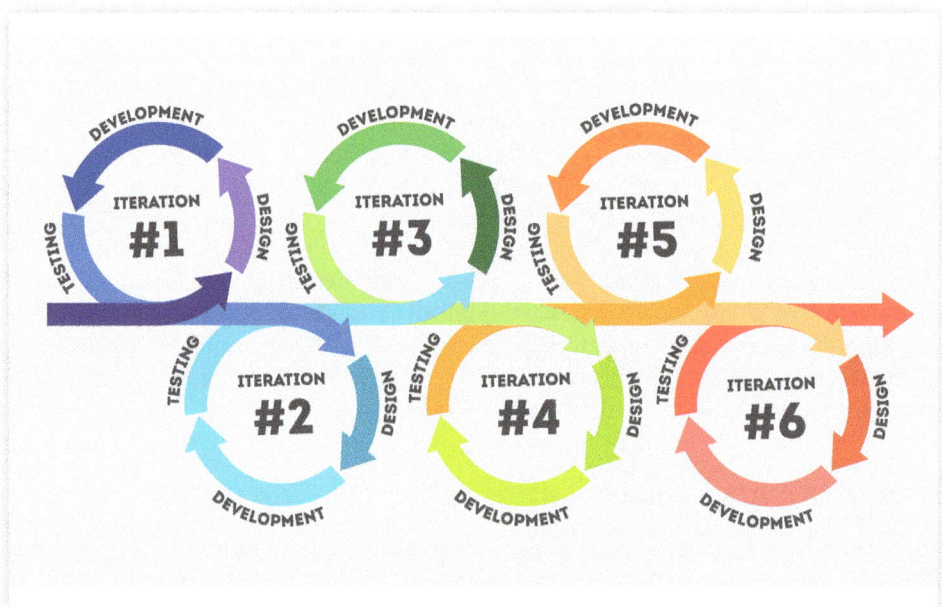

Because of their well-managed design frameworks, staged procedures were widely used for many years. These processes meticulously carry out a set of steps, are distinguished by limited iterations and strict reviews, and have a propensity to early specification freezing. Companies benefit from frozen specifications because they offer stability, sharpen product definitions, prevent scope creep, and lessen the need for midstream adjustments. When product cycles have consistent product specifications, strict quality requirements, and well-understood technology, as is frequently the case with product upgrades, staged processes function very well. Except for the comprehensive design stage, when multiple internal iterations must be completed before prototyping can start, stages are often distinct and sequential. Because it is one way, the staged process is sometimes referred to as the "waterfall" method. It is typically difficult or expensive to go back once a stage is finished [12].

Constraints and Boundary Conditions

Over the past ten years, there has been a steady rise in the application of nonlocal models in science and engineering. The motivation for and interest in this kind of simulation stems from the capacity of nonlocal theories to faithfully capture effects that are challenging or impossible to portray using local partial differential equation (PDE) models. For instance, a potential-based atomistic model accurately depicts material flaws like dislocations and interacting point flaws that are important in defining the elastic and plastic response of a material. The same is true for nonlocal continuum theories, which permit interactions at a distance without physical contact and may precisely resolve small-scale features like crack ends and dislocations. Examples include peridynamics and physics-based nonlocal elasticity.

The homogeneity of nonlinear damage models can potentially lead to the emergence of such models. Like how local response modifications to the Poisson-Boltzmann equation are qualitatively inaccurate, nonlocal electrostatic models have proven crucial in simulations of electrokinetic nanofluidic channels, for example. Nonlocal models are more accurate than typical PDEs but at the cost of a large increase in computing expense. Particularly, for many applications in science and engineering, a fully nonlocal simulation is still computationally unfeasible. Additionally, nonlocal models typically call for the use of volume constraints, which are trickier to define in real life than the local theory boundary requirement (Figure 6.8) [13].

FIGURE 6.8   We need tangible attributes to measure along the way.

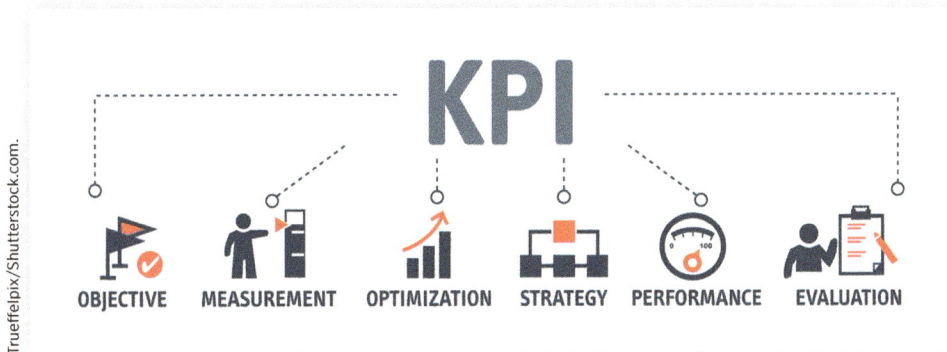

Trueffelpix/Shutterstock.com.

Product and Manufacturing Evaluation

Key Product Characteristics

Technical perfection is no longer sufficient to draw buyers to a product. Absolute errorlessness is a prerequisite. But how can sales be increased by getting buyers excited about new products? Making the best items for each client is the solution to this problem. All sensory impressions—visual, tactile, auditory, gustatory, and olfactory—contribute to the perceived quality of a product. Perceived quality adds consumer group-specific and subjectively perceptible qualities to the fundamental idea of quality. Unfortunately, there has not yet been a systematic method for bridging the gap between subjective consumer perception and objective product qualities. The manner of thinking in PD needs to adapt to handle these issues.

The technical quality of products has long been one of the success determinants for businesses in high-wage nations. The whole market segment now offers technologically superior items as a result of a focus on preventing failures. To keep or overtake the market, new success factors must be found. Many studies demonstrate that in addition to technical quality; subjective aspects including quality perception, brand image, price, and design affect customers' perceptions of quality. For instance, a study on over-the-counter items in the pharmacy branch reveals that the perceived product quality (including subjective and objective criteria) scored 33%, which is eleven percentage points higher than the brand as a customer influence factor.

Future items must not only reveal excellent quality but must also be recognized as high quality by the buyer to stand out from rivals. Production companies' development teams must concentrate on the so-called perceived quality, or customer quality perception. Customers divide product-related criteria into two categories: hard factors (objective quality criteria), which can be measured, and soft factors (subjective quality criteria), which are related to things like design or car configuration. As a result, the term "user-perceived quality" for technical products needs to be established and secure integration of emerging criteria for PD must be ensured. These activities need to be handled by being afterward integrated into all business operations [14].

Critical Characteristics

An evaluation method based on the preference ranking organization method for enrichment evaluation (PROMETHEE) is proposed to address the issue of how to construct quantitative evaluation index models that reflect the fundamental characteristics of reconfigurable manufacturing systems (RMS) and rank alternative reconfiguration schemes, which possess both advantages and disadvantages. The key characteristics of an RMS (scalability, convertibility, diagnosability, modularity, integrability, and customization) are quantitatively modeled based on an analysis of the reconfiguration of the reconfigurable machine components and manufacturing cells. The quantitative models are then used as the foundation for developing an RMS evaluation index system. The weights for these indices are assigned using the analytic hierarchy process [15].

Lead and Lag Indicators

In the engineering sector, knowledge management (KM) is gaining popularity. Companies are seeking measures to defend their investment in KM projects as they see increased investment. This study creates a methodology to evaluate the contribution of KM solutions to a company's corporate goals. Key performance indicators are used by the framework as lead indicators. At the strategic level, the lead indicators are created in accordance with the lag indicators. The framework is implemented in a company using a variety of templates.

To depict the performance of the company in relation to the business goals, lag indicators are created from the business objectives. They are lagging indicators that show how successfully a company's plan performed in the past. Additionally, they are generic in the sense that all businesses strive to advance in these areas. The lead indicators, which show the performance of the unique concerns of each organization, fill this gap. It is necessary to translate the organization's high-level strategic objectives and actions into steps that each person can execute to support the organization's objectives. But many businesses have had trouble breaking down high-level strategic initiatives, particularly nonfinancial ones. These lead indicators vary depending on the business's features and the domain in which they will be used. The percentage of goods for which the first design of a device fully satisfied the customer's functional specification was added as a lead indicator to the lag indicator in the BSC example of a PD department, where time to market was one of the primary outcome measures [16].

Functional Attributes

Many characteristics along the symbolic-functional continuum can serve as the foundation for brand positioning. While some brands may distinguish themselves largely through symbolic characteristics to establish, for instance, a prestige position, others may place more emphasis on functional distinctions that center on practical necessities. How customers regard their positioning features affects their choice of products and brand loyalty. This chapter adds to the findings from Wood [17], which identified brand loyalty among consumers aged 18 to 24 across six product categories and looked at the factors that influence brand choice [17]. This study sought to determine the level of brand loyalty demonstrated by 18- to 24-year-olds and whether there were any differences in loyalty according to the type of product. The age group was picked because, compared to other age groups, it is found to be less loyal. According to Wood, assuming low loyalty among this age group would be to disregard the intricacies of their purchasing behavior. Loyalty would also be influenced by the sort of product. There have been many studies that have examined the significance of product type. The products chosen for this study were toothpaste, toothbrushes, coffee, cereal for breakfast, sneakers, and jeans. The goods were chosen to largely mirror those picked by prior research focusing on product-specific loyalty and their usage by the age group under consideration.

The findings indicated that product type affects how loyal consumers are; hence, any findings from loyalty studies conducted on this age group should not be extrapolated to other products. In particular, the relevance of "quality" traits as a selection criterion is highlighted in this article discussion of the characteristics that affect the choice of product and brand within the food and toiletry product categories. Wood emphasized the significance of "quality" as a criterion for product selection. Additionally, since "quality" and "brand" positioning may both be distinguished along the functional/symbolic continuum, an examination into the quality factors that influence product choice is based on this association [17].

Performance Attributes

Consumer tastes are changing at faster rates now, and product lifecycles are getting shorter. Effective NPD is now more essential to business success than ever before. Many enterprises and academic fields, including marketing, technology, organizational theory, and engineering, have conducted more and more research on NPD. The key to successful NPD is developing products with core features that satisfy the needs of customers as well as other internal and external stakeholders, and auxiliary attributes that help differentiate products from one another. When creating a new product, stakeholder needs may be many and diverse and necessitate making trade-offs between competing needs. Customers may want ease of use and inexpensive prices, while the government and consumer advocacy groups may push for enhanced product safety. A technique known as "Design for X," where "X" stands for a specific collection of product attributes above and beyond fundamental functionality, such as "ease of use," "quality," or "safety," is used to select the right range of product features. During the past ten years, "design for environment," which is described as "a practice by which environmental issues are integrated into product and process engineering design procedures," has been a significant "X factor" [18].

Key Process Characteristic

A process model may be both prescriptive and descriptive, or it may combine elements of both. A descriptive process model documents unspoken information about how work is actually carried out. It makes an effort to highlight significant aspects of reality "as is." It is constructed deductively. A prescriptive process model, on the other hand, instructs employees on what to do and possibly even how to accomplish it. It is constructed deductively, sometimes using documentation from other projects or a third-party standard. A prescriptive process is a prescribed course of action that must be followed exactly. As the task becomes more routine, such as when we transition from PD projects to the traditional business procedures mentioned above, prescriptive processes become more appropriate.

There are several descriptive and prescriptive traits that many process models have in common. Of course, prescribing a procedure in the incorrect setting is risky. A process model should amass sufficient knowledge, experience, and accuracy before turning prescriptive to guarantee its viability and efficacy. Even yet, the PD environment is sufficiently dynamic to prevent any process model from ever becoming or remaining entirely accurate and complete. The main goal of this study is to develop descriptive process models that can be used to understand what and how work is done, but which are not always required to be followed precisely on a project. Project planning and execution are greatly aided by having a shared, accepted representation of a strategy and a network of commitments (described below) that are known to have worked in a somewhat similar circumstance. The process details also give the employees the ability to control themselves. Instead of creating a prescriptive process out of thin air, it is better for an organization to progressively transform elements of a descriptive process model into a prescriptive one.

Are standardized process models like the systems engineering (SE) "Vee" model and the spiral development model prescriptive or descriptive? They have elements of both. Many businesses use them as recommendations for their prescriptive procedures since they are high-level, broad descriptions of PD processes. Businesses may create tiered process models with descriptive lower levels and prescriptive higher levels. Unfortunately, several businesses are unaware of the difference, which leads to confusion and internal arguments about what constitutes formal, auditable, etc. process documentation. It is important to not undervalue the time and effort that is wasted as a result of huge enterprises [19].

Customer and Product Samples

Marketing

Therefore, it is thought that consumer and supplier involvement have a significant impact on the development of innovative products. The study of the precise aspects that can determine whether or not customers and suppliers would participate in NPD activities has received far less attention from researchers. One of the few studies which have looked at this topic discovered that elements like shared commitment, shared trust, shared adjustments, and shared relationship management affect customer involvement. This study builds on this line of thinking by paying close attention to the company's marketing plan. Investigating several key components of a marketing plan and determining how they affect customer involvement are the goals of the study. We see brand profiling as a key component of marketing strategy in this situation, along with product differentiation and competition focus. Additionally, we look at the function that particular investments play in facilitating consumer interaction and include them as a sign of commitment. Additionally, because most markets are growing more global and international, we place a strong emphasis on building relationships with foreign consumers. This is crucial for the success of our worldwide operations.

Customer Involvement in NPD According to the justification, such external communication increases the quantity and variety of information, which in turn improves the process quality. Direct communication with clients can be viewed as a high-bandwidth communication method. The ability to communicate complicated, ambiguous and unfamiliar information is made possible by informal, typically face-to-face communication, and it also gives room to take advantage of unexpected and surprising responses. First, we support their argument and contend that a key component of a product configuration is represented in the company's marketing plan. Second, including customers in the creation of new products calls for commitment from both the supplier and the customer to the partnership. The configuration is represented in the marketing strategy of the company.

NPD is a value-creation approach that requires actors to invest in certain assets committed to the relationship. To create value, NPD calls for actors to invest in certain resources that are dedicated to the connection. This encourages the growth of tight relationships that are marked by dedication and a long-term outlook. Thus, the amount of specific investments and the companies' marketing strategy can be seen as two key elements influencing whether or not customer participation will occur. In addition, the level of specific investments is likely to be impacted by the companies' marketing strategy. In the sections that follow, we will construct ideas about how various elements of a marketing strategy connect to consumer involvement in the creation of new products and particular investments. We concentrate on three crucial elements that make up a marketing plan. First, businesses frequently provide goods and services that clients perceive as distinct from those of rivals; this is what we refer to as "product differentiation." Second, organizations frequently employ commonplace strategies to alter their value propositions and improve their position in comparison to rivals; we refer to this as "competitor orientation." Third, companies frequently use various levels of brand profiling and firm reputation in marketing and sales, and we call this "brand profiling focus" [20].

Customer

Today's businesses must contend with competition daily, not just among themselves but also in every industry. To gain and keep their market share, businesses must be inventive and creative. Since businesses offer both their products and their values to their clients, the idea of value emerges first in this modern marketing period before the concept of the product. According to the theory of service-dominant logic (SDL), proposed by Vargo and Lusch, market actors who initially offered dominant logic (DL) in the forms of products and their attributive features are now trending toward SDL, putting more emphasis on their delivery value rather than just the products themselves [21].

Deregulation has led to a situation where a growing number of goods and services are comparable to, if not identical to, one another, underscoring the importance of a company's customer base. Customer retention is a difficult task that calls for a shift away from transactional marketing toward relationship marketing, which is extensively used nowadays and utilized to highlight a company's superior competitive position. People who wish to share information congregate on social media; it is a place to meet new people and engage in online interaction. Social Customer Relationship Management is the practice of using social media as a platform for businesses to control their online social interactions with customers.

The difference with the service concept put out by SDL is in how a business handles its customer interaction points by hoping that clients would recognize the presence of a moment of truth; if a business manages it properly, higher values can be generated. It was argued that these values originate more from the results of marketers' connections with customers through collaborative activities or co-creation value than they do from the final products.

Technical

Businesses are forced to create new items at an accelerated rate by fierce competition. Engineering teams are under a lot of pressure from this obligation to produce things faster while also producing better products. The idea of concurrent or simultaneous engineering has gained popularity as a result of these two difficulties faced by engineering organizations. Most literature on concurrent engineering cover how to design products using a successful multifunctional team strategy. The recommendation is to allow design engineers to collaborate closely with production engineers, field service engineers, and representatives of other parties involved in the creation and usage of the product to integrate the many challenges. There are several tales of small projects (five to ten employees) that used this "team integration" strategy successfully. This is what we refer to as concurrent engineering in tiny. It functions because small teams may collaborate closely, and difficult technological problems are exposed and handled through mutual understanding [22].

Evaluation Approaches

Virtual Reality and Augmented Reality

To examine a variety of options, VR and AR technologies have emerged as important in product realization. The study suggests a two-phased strategy, with the first phase consisting of a qualitative study to assess value stream collaborators' perspectives on the possible benefits and drawbacks of incorporating VR and AR into the NPD process. In the second phase, a quantitative survey was conducted to determine whether customers of apparel were aware of VR or AR applications, had opinions about them, and intended to use them in the future. Then in-depth interviews with industry professionals and 94 responses to a survey of UK consumers of clothes were used to compile the data. It has been determined that VR and AR technologies will make NPD successful in the garment sector by enabling quick replies to customers to improve the functionality of the new items.

VR and AR are now only a small part of applied PD, and their use in the garment sector is still in its infancy. To contribute to early investigations of such applications in this field, this study set out to design a collaborative NPD process that would enable VR and AR to make it easier for relevant partners to work on the same platform inside the garment industry. The successful application of digital technologies to suit the changing needs of the consumer has been assessed to depend on consumer engagement, personalization, and product visualization. As clothing needs have become more individualized in terms of aesthetics and functionality, digital technologies have the potential to improve PD capabilities in the manufacturing of apparel. Out of this, VR and AR technologies reveal potential applications in apparel, similar to other industries [23].

Decision Matrix (Pugh)

Global legislation has recently increased the demand for sustainable products, considering all economic, sociological, and environmental factors in addition to profit-making, consumer wants to fulfillment, and environmental impact reduction. The introduction of many sustainable product designs has resulted from the consideration of sustainability while creating product concepts, yet the sustainability considerations made when comparing various product configurations are quite constrained. In response, Hassan et al, suggests a methodical strategy to assess the sustainability of configuration design choices based on a weighted decision matrix and artificial neural network. When dealing with inconsistent data based on a point scale, a weighted decision matrix is utilized as a platform to construct ratings depending on the score supplied. In this step, configuration designs are rated on an interval scale and given a score using a nine-point Likert scale. This score is used to compare the configuration design to other options based on how well it meets the sustainability requirements. Then, an artificial neural network is used to combine the generated scores using a trained network to assess the sustainability performance as a single value, known as the Weighted Sustainability Score.

To fully explain the suggested strategy, a case study of an armed chair is done. Commercial software that is based on the environment validates the accuracy of the suggested approach in terms of the results for environmental evaluation. The findings show that, in rating the various part configurations with regard to environmental consideration, the suggested approach offers similar decision-making as the commercial software [24].

Market Testing

The activities necessary to test both the tangible product and the launch strategies in the target market are referred to as market testing. Launch budgeting refers to a budgeting task necessary to create, carry out, and keep track of a launch plan and tactics. The duties necessary to respond to the what, where, when, and why to launch questions are related to the launch strategy. Product techniques, distribution, price, and promotion are all part of the decisions made about the marketing mix that will be used to launch the new product. The four scales used to measure new product launch proficiency cover market research, launch budgeting, launch strategy, and launch tactics [25].

Technical Testing (Verification)

Choosing the best timing, frequency, and fidelity for sequential testing activities that are carried out to assess new product concepts and designs is the fundamental challenge in managing PD (Figure 6.9). In *Sequential Testing in Product Development*, the authors create a mathematical model that views testing as a process that uncovers issues with technical and customer-need-related issues. A model analysis yields a number of significant conclusions. First, the tension between several factors, such as the rising cost of redesign, the cost of a test as a function of fidelity, and the correlation between successive tests, must be balanced in optimal testing procedures. Second, a straightforward version of our approach yields an outcome similar to Economic Order Quality (EOQ), a company's optimal order quantity that meets demand while reducing cost: The square root of the ratio between preventable cost and test cost is the economic testing frequency, or ETF, which is the recommended number of tests. Third, the best testing approaches may be influenced by the link between successive tests. Budgets should be allocated to a few high-fidelity tests if consecutive tests are improving on one another. However, if tests are identifying problems independently of one another, it may be more effective for developers to conduct a greater number of lower-fidelity tests [26].

FIGURE 6.9 Effective technical testing of the product requires a variety of approaches and focus.

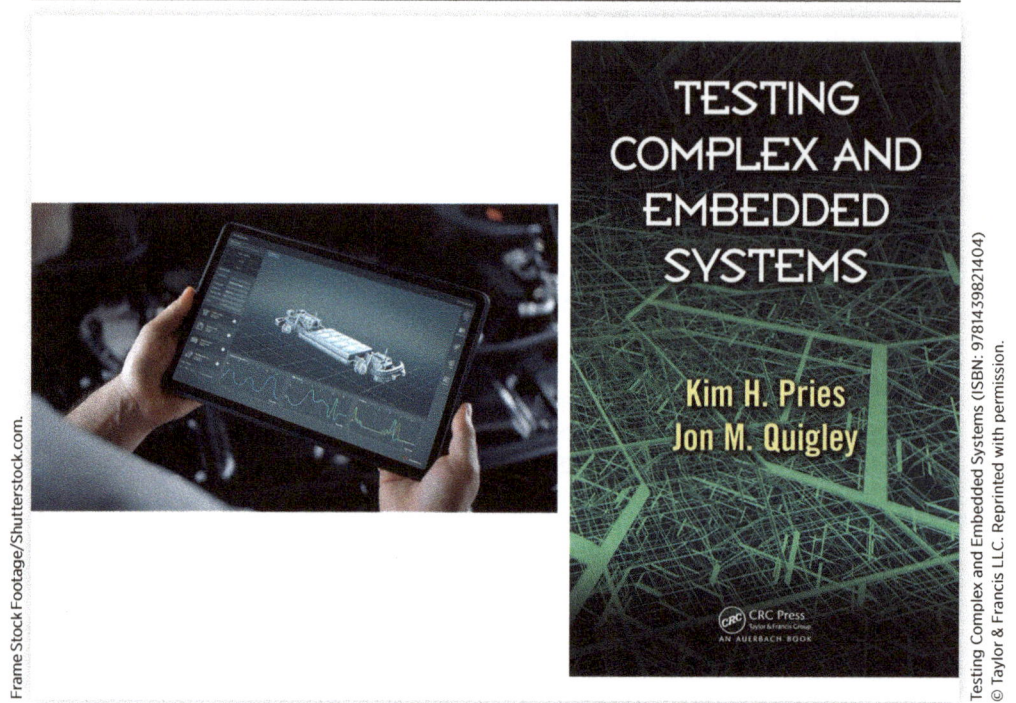

TESTING COMPLEX AND EMBEDDED SYSTEMS

Kim H. Pries
Jon M. Quigley

CRC Press
Taylor & Francis Group
AN AUERBACH BOOK

Frame Stock Footage/Shutterstock.com.

Testing Complex and Embedded Systems (ISBN: 9781439821404)
© Taylor & Francis LLC. Reprinted with permission.

Validation

Companies have concentrated on incorporating downstream PD processes into the design phases to enhance engineering product designs and decrease the number of issues that arise during a new product launch. Unfortunately, the tail end of PD has seen less of this resource integration to address issues, especially for complicated goods with many components, like an automotive body. Here, validation methods are used sequentially by manufacturing companies to first validate components, then subassemblies, and lastly the finished product. One trend has been to reduce or eliminate downstream assembly issues by tightening component tolerances. However, because of needless component rework, these tighter component requirements frequently cause timing delays and cost overruns. The use of "flexible criteria" and a strategy known as "functional construction" for complicated assemblies can greatly cut down on validation time and expenses while still maintaining end-product quality goals.

NPD is given more priority by manufacturers as product lifecycles get shorter. Costs and lead times have been decreased as a result of these efforts. Through techniques like concurrent engineering, design for manufacturability, and partnership sourcing, development resources have been more effectively incorporated into the product and process design phase, which has been a major driver of these advances. The backend of PD, however, has seen much less of this resource integration. Manufacturers frequently aim to carry out a sequential validation process during this step, whereby they first attempt to buy off or approve each individual component. Manufacturers confirm subassembly and final assembly procedures after components are approved. This sequential method is based on the fundamental tenet that the quality of the final product is maximized when the mean of each individual component dimension is generated according to the goal specification with the least amount of variance.

A sequential process validation strategy, where final assembly validation begins on time with all component quality standards met, is rarely used by manufacturers in practice. Instead, manufacturers begin assembly validation using unapproved components or postpone it as they wait for suppliers to address nonconforming faults or incorporate last-minute technical revisions. The Pareto principle is frequently applied to the costs connected with these delays, where revising 10–20% of component dimensions may account for 80–90% of the overall late time. When manufacturers find new assembly issues during final assembly validation, the effects of these delays frequently worsen, necessitating further component rework. Additionally, it has been questioned if it is important to meet the initial requirements when some of the nonconforming issues addressed during component validation are subsequently discovered to have a negligible effect on the final assembly. The number of issues that arise in the later stages of PD can be greatly reduced with greater resource integration during the design phase, yet issues still occur. Therefore, the efficiency of locating and fixing manufacturing validation issues becomes crucial to the overall costs and turnaround time for a new product launch.

Design of Experiments

The process variables used in the design of experiment (DOE) approach are first "screened" to see which ones are crucial to the result (excipients type, percentage, disintegration time, etc.). The second step is "optimization," in which the ideal values for the crucial variables are established. It entails using "mixture designs" to alter the composition of the mixture and investigating how these modifications may impact the qualities of the combination.

Advantages of DOE (Figure 6.10):

- Increased innovation as a result of process improvement.
- More effective manufacturing technology transfer.
- Fewer batch errors.
- A higher level of regulator confidence in durable products.

FIGURE 6.10   DOE is used to identify variable interactions to optimize quality.

Gorodenkoff/Shutterstock.com.

References

1. Tatikonda, M.V. and Rosenthal, S.R., "Successful Execution of Product Development Projects: Balancing Firmness and Flexibility in the Innovation Process," *Journal of Operations Management* 18, no. 4 (2000): 401-425.

2. Merminod, V. and Rowe, F., "How Does PLM Technology Support Knowledge Transfer and Translation in New Product Development? Transparency and Boundary Spanners in an International Context," *Information and Organization* 22, no. 4 (2012): 295-322.

3. Marle, F. and Gidel, T., "A Multi-Criteria Decision-Making Process for Project Risk Management Method Selection," *International Journal of Multicriteria Decision Making* 2, no. 2 (2012): 189-223.

4. Becker, M.C., Salvatore, P., and Zirpoli, F., "The Impact of Virtual Simulation Tools on Problem-Solving and New Product Development Organization," *Research Policy* 34, no. 9 (2005): 1305-1321.

5. Lewis, M.W., Welsh, M.A., Dehler, G.E., and Green, S.G., "Product Development Tensions: Exploring Contrasting Styles of Project Management," *The Academy of Management Journal* 45, no. 3 (2002): 546-564.

6. Petersen, K.J., Handfield, R.B., and Ragatz, G.L., "Supplier Integration into New Product Development: Coordinating Product, Process and Supply Chain Design," *Journal of Operations Management* 23, no. 3-4 (2005): 371-388.

7. Raudberget, D., "Practical Applications of Set-Based Concurrent Engineering in Industry," *Strojniski Vestnik* 56, no. 11 (2010): 685-695.

8. Crnkovic, I., Chaudron, M., and Larsson, S., "Component-Based Development Process and Component Lifecycle," Paper presented at in *the 2006 International Conference on Software Engineering Advances (ICSEA'06)*, Cavtat, Croatia, 2006.

9. Ayag, Z., "An Integrated Approach to Evaluating Conceptual Design Alternatives in a New Product Development Environment," *International Journal of Production Research* 43, no. 4 (2005): 687-713.

10. Michaelis, M.T., Johannesson, H., and ElMaraghy, H.A., "Function and Process Modeling for Integrated Product and Manufacturing System Platforms," *Journal of Manufacturing Systems* 36 (2015): 203-215.

11. Katila, R. and Mang, P.Y., "Exploiting Technological Opportunities: The Timing of Collaborations," *Research Policy* 32, no. 2 (2003): 317-332.

12. Unger, D. and Eppinger, S., "Improving Product Development Process Design: A Method for Managing Information Flows, Risks, and Iterations," *Journal of Engineering Design* 22, no. 10 (2011): 689-699, doi:10.1080/09544828.2010.524886.

13. D'Elia, M., Perego, M., Bochev, P., and Littlewood, D., "A Coupling Strategy for Nonlocal and Local Diffusion Models with Mixed Volume Constraints and Boundary Conditions," *Computers & Mathematics with Applications* 71, no. 11 (2016): 2218-2230.

14. Lieb, H., Quattelbaum, B., and Schmitt, R., "Perceived Quality as a Key Factor for Strategic Change in Product Development," Paper presented at in *the 2008 IEEE International Engineering Management Conference*, Estoril, Portugal, 2008.

15. Wang, G.X., Huang, S.H., Yan, Y., and Du, J.J., "Reconfiguration Schemes Evaluation Based on Preference Ranking of Key Characteristics of Reconfigurable Manufacturing Systems," *The International Journal of Advanced Manufacturing Technology* 89, no. 5 (2017): 2231-2249.

16. del-Rey-Chamorro, F.M., Roy, R., Van Wegen, B., and Steele, A., "A Framework to Create Key Performance Indicators for Knowledge Management Solutions," *Journal of Knowledge Management* 7, no. 2 (2003): 46-62, doi:https://doi.org/10.1108/13673270310477289.

17. Wood, L., "Functional and Symbolic Attributes of Product Selection," *British Food Journal* 109, no. 2 (2007): 108-118, doi:https://doi.org/10.1108/00070700710725482.

18. Pujari, D., Wright, G., and Peattie, K., "Green and Competitive: Influences on Environmental New Product Development Performance," *Journal of Business Research* 56, no. 8 (2003): 657-671.

19. Rezayat, M., "Knowledge-Based Product Development Using XML and KCs," *Computer-Aided Design* 32, no. 5-6 (2000): 299-309.

20. Svendsen, M.F., Haugland, S.A., Grønhaug, K., and Hammervoll, T., "Marketing Strategy and Customer Involvement in Product Development," *European Journal of Marketing* 45, no. 4 (2011): 513-530, doi:https://doi.org/10.1108/03090561111111316.

21. Vargo, S.L. and Lusch, R.F., "The Four Service Marketing Myths: Remnants of a Goods-Based, Manufacturing Model," *Journal of Service Research* 6, no. 4 (2004): 324-335.

22. Eppinger, S.D., Whitney, D.E., Smith, R.P., and Gebala, D.A., "A Model-Based Method for Organizing Tasks in Product Development," *Research in Engineering Design* 6, no. 1 (1994): 1-13.

23. De Silva, R., Rupasinghe, T.D., and Apeagyei, P., "A Collaborative Apparel New Product Development Process Model Using Virtual Reality and Augmented Reality Technologies as Enablers," *International Journal of Fashion Design, Technology and Education* 12, no. 1 (2019): 1-11.

24. Hassan, M.F., Saman, M.Z.M., Sharif, S., and Omar, B., "Sustainability Evaluation of Alternative Part Configurations in Product Design: Weighted Decision Matrix and Artificial Neural Network Approach," *Clean Technologies and Environmental Policy* 18, no. 1 (2016): 63-79.

25. Langerak, F., Hultink, E.J., and Robben, H.S.J., "The Impact of Market Orientation, Product Advantage, and Launch Proficiency on New Product Performance and Organizational Performance," *Journal of Product Innovation Management* 21, no. 2 (2004): 79-94, doi:https://doi.org/10.1111/j.0737-6782.2004.00059.x.

26. Thomke, S. and Bell, D.E., "Sequential Testing in Product Development," *Management Science* 47, no. 2 (2001): 308-323, doi:10.1287/mnsc.47.2.308.9838.

7

Benefits of Multiuse and Reuse

Modularity

What Is Modularity?

Rapid innovation and mass customization present a new competitive advantage in today's business world. As a result, it is transforming how many organizations conduct business in the fiercely competitive global economy (Figure 7.1).

FIGURE 7.1 Modular designs can help with manufacturing as well as design extension to multiple iterations and market adaptations.

To meet a wide range of client requirements, businesses have successfully used strategies to create and develop product families based on shared product platforms. The product family is a term used to refer to a collection of similar goods that serve several different market segments and have identical characteristics, components, and subsystems.

Increasing commonality and standardization within a product family is a primary objective during the design process for product families. Modularity is the most well-known strategy for the efficient design of product families. Think Lego blocks when you think of modular systems (Figure 7.2).

FIGURE 7.2 Modular systems allow for complex things to be built from basic building blocks.

The term "modularity" refers to the process by which a product is divided into several different assemblies, each designed to carry out a particular task. Even so, the objective of the product is accomplished by the interaction of all the product assemblies (or modules).

Baldwin and Clark [1] describes the concept of modularity as an organizational technique that is used to manage complicated products and processes effectively. As a result, we save time to market, material costs because of the economy of scale, and opportunities for improved quality. Modularity is a subfield of a broad topic in theoretical product development called product architecture. Creating modularity across design solutions suitable for various product configurations is the topic of discussion in this chapter.

Historically, companies have developed products and solutions by taking a dedicated project team focused on specific project requirements. Modularity provides more flexibility to the companies as they transition toward scaling their product to multiple variants or configurations and are challenged with R&D expenses.

Common Solutions

At the industry level, product and organizational modularity are credited with the development of platforms utilized by a complete industry and, as a result, industry structures. These structures enable teams from different departments to work independently on loosely related challenges [2]. In addition, these arrangements permit teams from several departments to work independently to find standard solutions for different challenges.

Appropriately applied modularization of product structures can serve as a technique to give the customer-required diversity while reusing subsolutions across several products to enhance time to market and preserve predictable product quality.

Similar Functionality

Through coordination with the stage of backend product realization, the second phase of detailed platform development attempts to improve engineering efficiency by increasing the degree to which commonalities are shared. For the purpose of capturing and resolving the trade-off between the commonality configuration and the individual product performance, a method of optimization that is tractable is adopted.

Shared across Multiple Configurations

To define portfolio modules, product functions can be shared across multiple configurations. Functional groups with similar flows and functions that appear more than once in a full portfolio function structure should be put together into a single module. Then, this module can be used for other products in the lineup. To achieve these, changes are necessary for the development process toward modularity (Figure 7.3).

- Reuse components wherever possible
- Product road map aligned with future strategies
- Decision-making process

FIGURE 7.3 What are carry-over parts, and differentiation parts and functions?

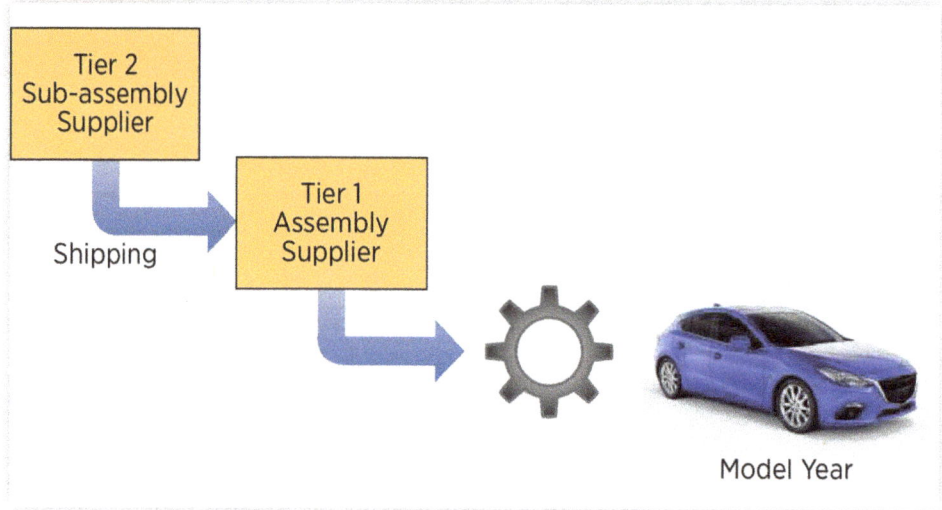

Companies can set up a new cross-functional organization that focuses on a modularity steering committee to analyze projects and use a disciplined process to determine which strategies can be developed as modular solutions for the projects. Instead of having a fixed product platform on which different products can be made by swapping out other add-on modules, this method lets the platform be one of several different sizes or types. We start by making function structures for each product in the portfolio.

Each structure is based on a different physical principle that is at the heart of the same technology. A different function structure system can be used for each physical principle that is being looked at. The function structures are compared to see which modules are the same and which ones are different. Then, rules of product modularity, such as dominant flow, branching, and conversion, are used to determine what other modules might be possible. Any consistent set of modularity rules that are followed defines a portfolio architecture that is possible. Finally, each portfolio architecture is shown by a modularity matrix of functions versus products (Table 7.1). The matrix shows how many functions are shared and how many are unique. This method gives a systematic way to come up with possible portfolio architectures and helps the design team talk about their ideas.

TABLE 7.1 Example of a modularity matrix.

Functions	Product features			
	Product Variant 1	Product Variant 2	Product Variant 3	Product Variant 4
Function A	X		X	X
Function B	X	X	X	X
Function C	X	X	X	
Function D	X	X		X
Function E		X		

Benefits of Modularity

When appropriately used, modularity allows businesses to achieve a number of strategically significant benefits in the process of competing in product markets. Some of the benefits of using modular designs in the process of product development include the following.

Improved Product Performance

Product performance can be enhanced by continuously improving the design modules that serve as the foundation for product development based on field-tested product data. By adopting modularity in the design of product family architectures, enterprises can offset the detrimental impact of product variation on operational performance, according to operations management research.

A modular product design can be technically partitioned in such a way that each product feature or functionality that is believed to be a big part of what makes the product different is either in a single component or in a group of components that make up a subsystem.

This can be done so that each function or feature of a product stays separate from the others. After that, multiple versions of functional components (or subsystems) can be added to the modular architecture to make different product versions with different combinations of functionality, features, and performance levels based on the components.

Reduce Product Variability

Modularity reduces product variability by increasing production consistency. The ability to produce more consistent output enhances product performance and may minimize manufacturing costs. By isolating the properties of a product responsible for driving performance into separate modules, modularization makes it possible to break the direct connection between product variation and the expense of complexity. By doing things in this manner, organizations can maximize the number of possible variations for a specific module without creating variations of anything else. Reducing product variability directly impacts various elements of the company's business, such as improving productivity and reducing labor and expenses (Figure 7.4).

FIGURE 7.4 The product variation requirements are influenced by many things.

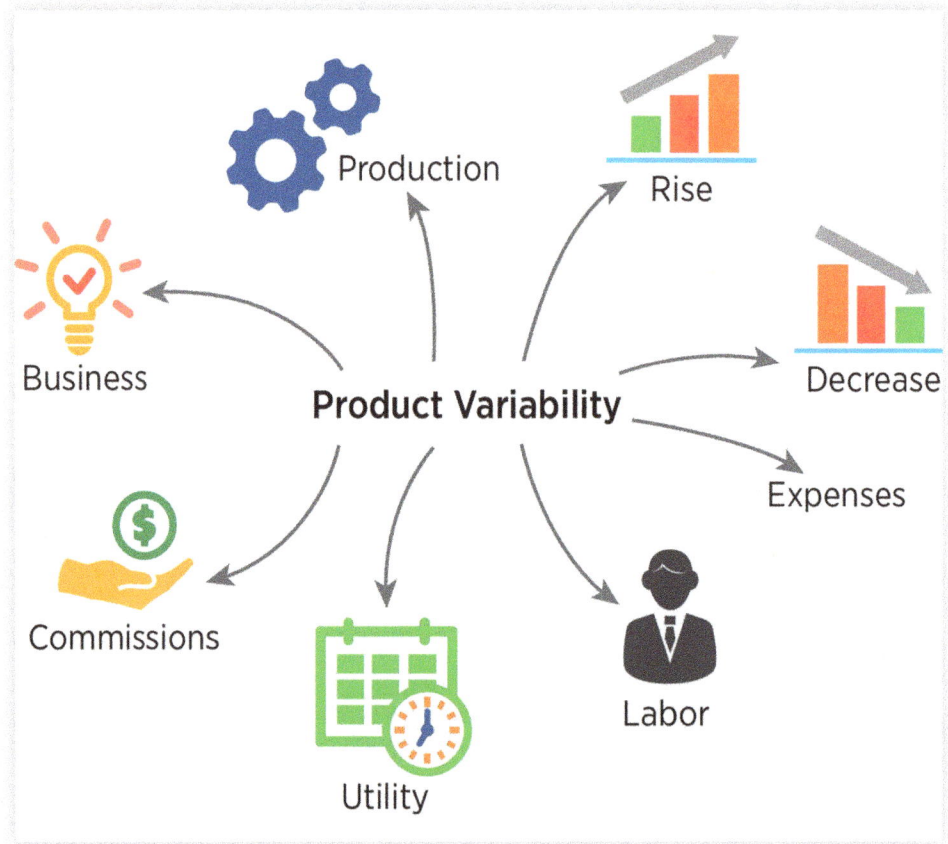

Increase Profitability

Sanderson and Uzumeri [3] and Henderson and Clark [4] found that using product platforms has given organizations a competitive advantage regarding the number of items they may offer and their profitability. Modularity increases the profitability of products in many ways. One of the ways is that it reduces time to market, increasing the chances of an organization gaining a competitive advantage. In addition, it reduces the complexity of product offering leading to improvement in the manufacturing processes and product quality.

Additionally, modularity increases product profitability with excellent product performance. Since different consumers measure performance differently, modularity is part of catering to individual customer needs, thus bringing value to the products from every angle.

Decreases Product Maintenance

Sony's Walkman is a great example of how product platforms, which are a type of modularity, can be used to build a whole family of products that do not require much maintenance. In the 1980s, Sony released more than 250 unique models in the US market alone, all of which were based on one of only three distinct platforms [3]. Most modifications to the models consisted of making subtle adjustments to the characteristics, packaging, and overall appearance of the products.

Ease of Manufacturability and Serviceability

When producing a product, modularity enables the procedures of developing components for the design to be partitioned into more minor activities. This makes the overall manufacturing process more efficient and much faster.

It is additionally easy to provide product serviceability and faster technological upgrade. However, the most important thing to remember is that modifications made to one product component will only affect specific other components. For instance, this was an essential characteristic for Sony to meet before it could launch later Walkman versions in the 1980s. Furthermore, modularity makes it simple to produce product designs ready to incorporate cutting-edge components as they become commercially accessible over the product lifetime (Figure 7.5).

FIGURE 7.5   Aftermarket serviceability is an objective of the development effort.

Nomad_Soul/Shutterstock.com.

Modularity is one of many ways how an organization establishes flexibility. According to Ethiraj and Levinthal [5], it has been suggested that modularity might be used to achieve the flexibility to better handle complexity and market unpredictability. This makes it easy for organizations to meet long-term industry demands.

Managing market unpredictability requires the utilization of modular architectures. For example, when managing the irreducible uncertainty regarding which product variations customers will want in the future, a modular architecture may be designed to accommodate a wide variety of product variations. This is done to mitigate future consumer preferences that are unsure of being known [6].

A corporation's ability to quickly and cheaply create new product variants generated from diverse combinations of current or new modules is made possible by flexible product design, enabling the company to respond to shifting consumer preferences and technological developments. As a result, an organization can meet long-term industry demands (Figure 7.6).

FIGURE 7.6   Interchangeable hardware and software modules extend the product range.

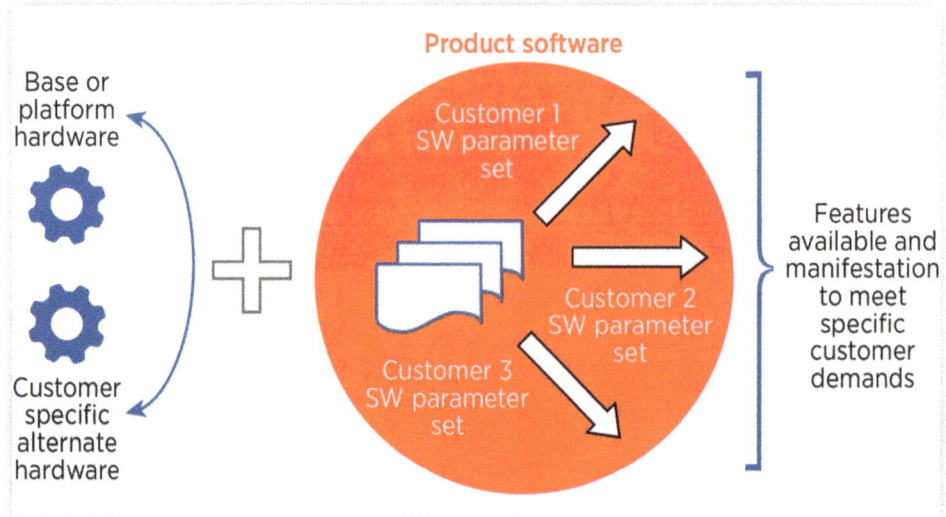

Software Modularity

Software products or software components of a product also benefit from modularity. It can be easier to implement and maintain with a measure of discipline associated with building individual software modules. In addition, we can use parameters within the software build to enable certain features related to a specific customer incarnation of the product.

Applications of Modularity

Any Industry

The application of a modular system is a mindset. Modularity is applicable across all industries, pushing the limits of what can be accomplished with product platforms and shared product components.

Where There Is High Demand for Product Volumes

Many fast-paced industries have been making their products in a modular way for a long time. But more and more of them are now ready to take modularity to the design stage. Modularity is most common in industries with high demand for product volumes, like the technological and car production industries.

Significant Product Variability Is Involved

To promote commonality within products without sacrificing variability, modularization is used. For example, standardized interfaces and space reserves make it possible for several module variations to be used interchangeably across a wide variety of products. As a product platform, organizations can only sell product variations that share the technical compatibility of modules with customers.

When making software systems, developers think about different development levels, but not just in a closed-world scenario. They also think about important things like nonfunctional properties and the specification of multi product line (MPL) behavior, which goes beyond a closed-world scenario.

It is argued that the right combination and communication between the software product lines of an MPL must be ensured on several levels related to the MPL development process. These levels include modeling variability, nonfunctional MPLs, implementation, and specification.

The modular system is handled as a whole from the point of view of product management to ensure that specific technological development projects for each module variant do not compromise the value created by the modular system.

Customer Requirements Are Rapidly Changing

In this day and age of rapid globalization and strong consumer demand, the production sector faces a significant issue in meeting the growing needs for customized goods that align with industry trends, branding strategy, and customer requirements.

A modular approach to product development not only helps cut down on the time needed to develop a product but also enables production to tailor products to the specific requirements of individual customers effectively. In addition, doing so establishes a distinctive brand identity and cuts the amount spent on product development.

Modularization of product platforms presents OEMs with an excellent opportunity to establish platform strategies that can be used by many brands and the capacity to tailor goods to specific customer needs. Furthermore, amid the current recovery from the Covid-19 pandemic, product manufacturers will need to reevaluate their production strategies and emphasize adopting a platform mentality to offer goods that are in line with customers' expectations.

Road Map and Strategy

Product platforms enable the rapid and cost-effective development of new high-value goods through the use of standardized modules and differentiation through the use of other modules. Because it is not essential to repeatedly go through the design process, a brand image may be developed while still bringing a huge quantity of items onto the market at a lower cost. The increasingly complex and globalized value chains for future technologies, which are increasingly cross-sectoral in nature or result in completely new industrial products and applications, systematic linkages between roadmaps with different application foci and from various industries must be developed and established. Road mapping projects need to be more modular as a result of this strategy [7].

To accomplish these goals, it is essential to first create a platform road map and a strategy, then carry out the analysis, and finish by designing the products.

Modularity Can Be Successful if Long-Term Strategy Is Clearly Defined

As a result of manufacturers being more conscious of the ever-changing client expectations, product sustainability is now being taken into consideration throughout the design process of new goods. Consequently, design for social sustainability of products has emerged as a new product design direction.

There is a need for the administration to make it clear what is expected of products. In addition, business models should be set up on a national level (at least) to develop a long-term strategy to speed up the adaptation and use of products and services that are more environmentally friendly. In this way, social institutions and businesses can work together to reach their future goals in product development.

Plan for Future Solutions with a Common Strategy

Because of growing awareness of resource waste and environmental contamination over the past two decades, sustainability has drawn increased attention. As a result, many studies have also been done to investigate sustainability concerns with product design. However, the ultimate aim of sustainable NPD is to provide for the requirements of the present without sacrificing those of future generations. Therefore, integrating crucial sustainability strategic factors is necessary to achieve this goal. Therefore, product sustainability over its full life-cycle is a pressing issue for product development companies, and one of many societal challenges designers in the field will need to address by creating a common strategy.

Reduces R&D Expense and Time and Improves Quality

To stay competitive in the market, one of the most efficient methods to achieve the aim of product diversity while simultaneously reducing the R&D cost of development is to design products with a modular structure for its component parts (Figure 7.7) [8].

FIGURE 7.7 Modular approaches can positively impact the quality, cost, and time to deliver.

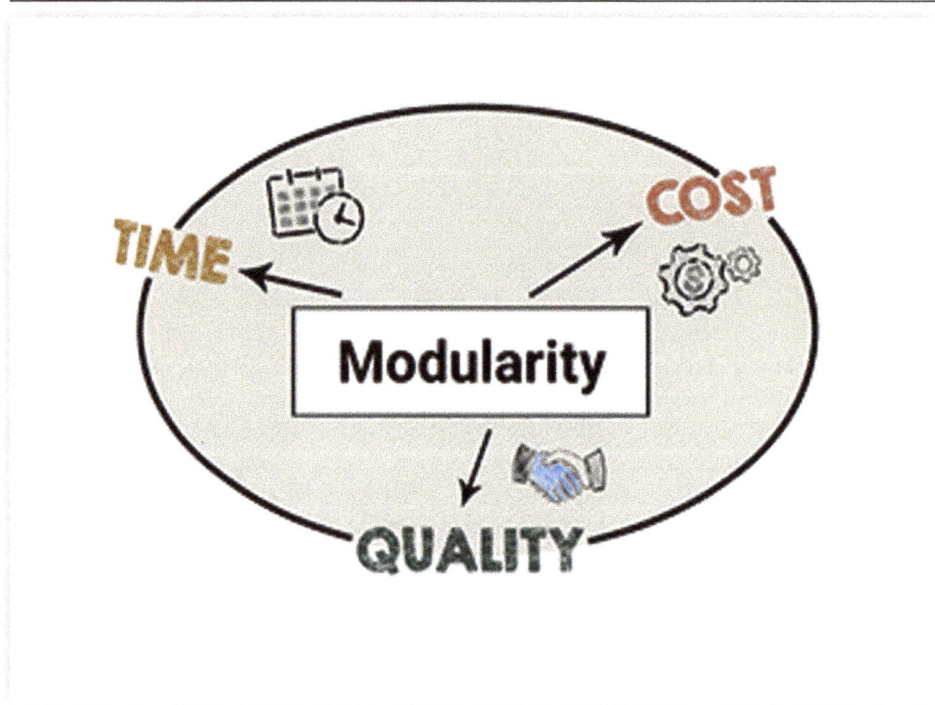

stoatphoto/Shutterstock.com.

The use of platform and module techniques can introduce the strategy to cut down on the number of product iterations, which in turn enables businesses to save money on R&D and reduce time to market for introduction. Moreover, standard modular designs and interfaces between modules can also make it simpler for engineers to implement incremental enhancements to particular modules without necessitating costly rework elsewhere in the project design.

In addition, considerable reductions in the time needed to get a product to market makes it possible for businesses to pool their R&D resources and create genuine innovation road maps rather than reengineering many variants of the same solution. They additionally assist teams in developing more compelling business cases for extra product versions. The use of platform and module solutions can result in a variety of potential benefits, including the apparent reduction in the cost of materials attained through increased or bundled purchases.

Establishing Requirements and Boundary Conditions

The first step to solving a problem successfully is clearly defining the boundaries and establishing the requirements necessary. From experience, the environment in which the product will be used is seldom easily understood. The same case applies to the development of product platforms. Since product development involves solving a problem in the marketplace, we utilize boundary conditions as a starting point for problem-solving and establish rules for what is and is not allowed as we work toward a solution.

Simulation is absolutely brilliant, if you understand your boundary conditions.—Perkins

Secure Early Requirements

After narrowing the portfolio to a physical concept, modularization considerations can be taken. At this phase, there is a lot of leeway in deciding how a product family should be built. When deciding which product partitioning modules will be used throughout a product family, a system engineer must take into account a number of factors, the most important of which are outlined below. Therefore, when making early decisions about system architecture, these four types of goals should be taken into account.

The first is traditional market variance, which looks at how much variety each customer needs, based on how different they are from each other.

The second is usage variation, or how a person who buys a product wants to use it in different ways after the purchase. This difference is often overlooked in market science literature and research, but it is important for figuring out what kinds of products are needed, such as multiple fixed products, swappable modules on a standard interface, or a platform that is easy to change.

The third factor is how fast the different parts of technology change before a new product design is needed.

The last type of influence is what we call "Design for X," which means how design, production, supply, and lifecycle criteria affect how products are split up.

By considering these four influences, an organization can come up with product requirements before beginning the development process.

Customer Expectations Are Clearly Defined

The first step in developing a platform is to investigate unmet customer demands by monitoring a representative sample of the target market to identify its members' preferences and expectations in terms of the essential features of the platform. The demand for an entire product family can be estimated using the functional view, which organizes customer needs and expectations. They are compared to the design parameters from a technical perspective, and then modules are derived from there. Otto and Wood [9] describe a flow-based approach for developing a function structure for any given product. A flow is identified for each client's need.

This flow is then tracked through the product as a series of subfunctions that change the flow, just as it would in usage. These independent chains are then combined to form a complete function structure network that meets the customer's needs to the greatest extent possible.

Boundary Conditions Defined to Secure Long-Term Product Evolution

When designing a new product or service, most organizations try to avoid or forget to think of the boundary conditions of the products. The ideas are based on what a product should do, but not what it is not meant to do or what the product should not do. Boundary conditions are essential for any team (or person) to remember before starting a new project. Boundary conditions let engineers convey NPD package support assumptions.

First, we identify the boundaries of the product family, which should share the same modules, and decide which underlying technologies should be used. After that, individual conceptual designs are created for each product. A product has a wide range of applications; many ideas exist.

Align Requirements and Boundary Conditions with X-Functional Network

Once the boundary conditions and product requirements are clearly defined, every product concept is used to create functional structures. Combining these function structures for every idea produces a large family function structure. The family function structure shows how all the family's product functions interact.

The modularity matrix, which contrasts the family's functions and products, contains the main details for the functions as they are used in each product. The architecting decision space is made up of the individual components of the matrix.

Manufacturability and Serviceability

Companies focus on product characteristics to improve NPD. Effective NPD must go beyond competitive features to fulfill after-sales service. After-sales service refers to all the events that occur during the use cycle of a product, from purchase to the point where customers drive the product out of market.

Customers need maintenance and repair from manufacturers to get the most out of their products. Design for Service or Support (DFS) improves client satisfaction and company income. DFS additionally enhances the manufacturability (the ease of manufacturing) of a product and its serviceability (how easy it is to service a product).

Define Manufacturability and Serviceability Requirements

The usage cycle begins with delivery, but manufacturing in the production facility significantly impacts serviceability. DFM is a commonly used technique in product development that reduces manufacturing costs [10]. It investigates production requirements in manufacturing throughout the design process, allowing the development of products that are easy to service and economical to build. There are many DFM methodologies, but they all examine manufacturing requirements during the design phase and provide quantitative targets for manufacturability.

For instance, methods can be employed to encourage design simplification and to assess how easy it will be to build a product. This may result in fewer modules, resulting in lower material costs and quicker assembly. One prominent method employs quantitative scores depending on the number and type of components [11].

Perform Early Design for Assembly and Design for Serviceability

Using DFM as a framework, it is important to establish priorities for the product and its serviceability and supportability during the design phase. Not only is there a similarity between DFS and DFM, but the two are also more closely related. Companies that have used DFM have often found that they also need to think about service and support issues because there may be trade-offs. This is because manufacturing can have goals against support and service [12]. For example, the manufacturing purpose is to reduce the cost of factory assembly, which may create a simple product but be difficult to disassemble and reassemble at the client site. Hence, it is crucial to perform design for product assembly as well as the design for serviceability early in the development process.

Common Installation Strategy Critical to Have a Modular Product

For a modular product, a standardized method of assembly is essential. A module is made up of many different parts combined into one unit.

However, engineers do not just put the pieces together. They aim to design a beneficial and valuable system for customers. This means that the parts are supposed to communicate with one another. Therefore, the techniques should interact more. Accordingly, engineers must manage the confluence between design and performance. The module strategy entails developing module capability as a systems integrator. When specific system components are created, tested, and supplied by various organizations, such cooperation is very crucial.

Setting the project objectives, developing a common installation strategy, directing its execution, assessing the results, and prescribing the required corrective actions to keep it on track are all important aspects of successful modular product development [13].

Modular Assembly Sequence Defined

Structures for manufacturing and assembly are defined. It is also necessary to identify and depict the general production and assembly sequence. A value stream analysis can also be performed. The foundation for further process analysis is the general manufacturing and assembly sequence.

The different components (parts) of the system (product) cannot be engineered independently of one another and then only put together to create a functional system. To ensure that the interactions and interfaces between system parts are compatible and mutually supportive, systems engineers must direct and coordinate the design of each piece as necessary.

It is initially necessary to build a generic product sequence based on the now available items. To assign subassemblies and components to the various levels, several aggregation levels must be established.

A hierarchical structure is created by joining the components and subassemblies. Depending on the complexity of the product and the degree of added value, different products may have a variable number of aggregation levels.

VR/AR Tools Used for V&V

Technologies already have a significant impact on the process of product development and will probably continue to have the same impact in the future. This has to do with how items are developed and produced as well as the technologies employed in those products.

Digitalization is one area of technology in particular that has had a big impact on both products and the manufacturing of those products. The expert group pointed out a few new skills that are appearing in this field that include the use of VR/AR tools to verify and validate products.

Digital tools and computer technology have enabled a greater number of iterations and consequently verification and optimization of product level. The progress is especially strong in certain fields, which is the classic "islands of automation" dilemma.

Even though more sophisticated tools for integrated multidisciplinary simulation are required for the V&V of future products, AR/VR technology may solve certain facets of the present-day product manufacturing.

The advent of VR/AR technologies has enabled the advancement of learning techniques in experimental ways.

Problem-Solving

Modularity Provides Opportunity to Troubleshoot Issues Quickly

An ideal modular product design consists of different products partitioned into practical and useful modules. As a result, smaller modules are easy to navigate through and spot problems, making troubleshooting and maintenance easy.

The effectively constructed modules can be easily updated on regular time cycles, some can be manufactured in several levels to offer a wide range of market options, some can be quickly removed when they wear out, and others can be easily exchanged to get additional functionality.

These benefits of successful product modularity are increased when identical modules are utilized in many products.

Most design issues can be divided into several manageable, easier-to-solve issues. Sometimes difficult problems can be broken down into simpler ones, and tiny adjustments to the solution of one subproblem might have a ripple effect on the solutions of other subproblems.

This indicates that functionally dependent subproblems have emerged as a result of the decomposition. The goal of modularity is to minimize interaction or interdependence between subissues by breaking the larger problem down into smaller, functionally independent problems.

As a result, altering the solution to one problem might only make modest changes to other problems, or it might have no impact at all.

Six Sigma Approach (Repeatability of the Problem because of Common Solution)

Six Sigma is a concept and approach for quality management that focuses on enhancing the quality of processes, goods, and services (Figure 7.8). Six Sigma is based on the principle of eliminating flaws by lessening the amount of variation present in all of a company's operations and products.

FIGURE 7.8 Six Sigma approaches to the product and process.

Trueffelpix/Shutterstock.com.

Six Sigma professionals typically employ the DMAIC technique. The five major steps of the process are defined by the acronym DMAIC, which stands for Define, Measure, Analyze, Improve, and Control.

- **Define:** At this stage, whatever it is that needs to be achieved is clearly defined. This stage focuses on the customer needs and objectives and goals are also created at this stage.

- **Measure:** The objective of the Measure phase is to collect as much information as possible to provide light on the current state of affairs as the team builds a process baseline. Various inputs and outputs are defined, and baseline levels are monitored prior to the implementation of improvement activities. In the Measure phase, the process sigma and process capability are determined for the defined processes. Throughout the process improvement program, data are accumulated and shown graphically to facilitate comprehension by all interested parties.

- **Analyze:** Key factors that affect the output are recognized during the Analyze phase of the DMAIC process, and the underlying reasons for any loss in productivity or quality loss are found.

- **Improve:** The objective of the Improve phase is to find answers that have been discussed, test them, and then develop a pilot "initial solution" that can be put into action. The emphasis during this phase is also on figuring out the ideal combinations of input parameters for the different outputs. When an initial solution is operational, the next phase is to iteratively improve it through kaizen events, employing the expertise of the process owners and operators. A cost-benefit analysis is then carried out to see how much progress the team has been able to capture after the process has stabilized. Kaizen events or kaizen blitzes, also known as Rapid Improvement Events, the execution of the first and the revised solution, and the monitoring of the impact of changes are all carried out during the Improve phase.

- **Control:** The project has been completed, closed out, and turned over to the day-to-day operators during the Control phase of the DMAIC process. Additionally, at this phase, the mechanisms and structures required to keep the process running at peak efficiency should be developed. Plans should be created and put into action to maintain the new process through ongoing observation and management. Plans for how the process team will use the process metrics as a new baseline will be implemented during the Control phase. They must then design and put into place a continuous improvement system. The assignment of duty and accountability for the process should be completed during this phase, and a schedule for routine process monitoring should be developed.

Traceability of the Issues Can Be Achieved with a High Degree of Accuracy

It is possible to expand the potential for innovation by supplying designers with relevant information regarding a product or assembly, such as the required function, the client requirement, any spatial limitations, and ways that have been effectively employed in the past.

This information aims to achieve traceability of product information to provide designers with the ability to make decisions regarding how previous designs may satisfy similar specifications of new products. Additionally, it helps the designers to spot problems that arise while coming up with proper problem-solving techniques.

This, in turn, enables the utilization of the existing product knowledge and shortens the amount of time needed to bring a product to market.

With modularity, intricate artifacts may be broken down into simpler, more independent modules. By allowing for tests to be done independently on individual modules rather than on the overall artifact, modularity reduces complexity and speeds up development [2].

Improved Quality

What is quality? By definition, quality is the standard by which something is judged based on how it compares to other things of the same kind. The quality of the product is one of the traditional competitive characteristics that plays a significant part in the process of developing successful products.

A variety of manufacturing strategies can be used to increase quality; nevertheless, modularity is the element that indicates the complexity of products as well as their potential to evolve.

High Quality Is Achieved with Modularity

The heart of a high-quality product is modularity. Some experts say that a product design makes up about half of its quality.

According to the findings of a study conducted by General Motors, employing a modular design and the uniformity that comes along with it leads to an improvement in the overall product quality [14]. Standardization, according to Kusiak [15], not only improves the quality of a good or service but also results in increased revenue for the business or organization that implemented it.

Fisher et al. [16] looked into this and came to the conclusion that there were several reasons why standardization improved quality. One was the effects of the learning curve that came from making more of the same parts or subassemblies.

Because a subassembly or component is utilized in a variety of products, its performance must meet the design standards for the most severe application. This indicates that its capabilities transcend all but the most rigorous application constraints.

Feitzinger and Lee [17] said that product modularity improves quality because problems can be isolated to specific modules, which makes it easier to fix them in a targeted way. Another thing that product modularity has to do with quality is how it affects how consumers see quality.

Onkvisit and Shaw [18] say that standardization gives a company a clearer image, which can affect how consumers see the quality of the product.

Reduced Product Variability Helps with Improved Quality

The decrease of variability in product qualities is a key objective of quality improvement in product development. In the product development process, the difference between the quality measure that is produced and the target quality measure is called "variability." When there is a high degree of variability, there is either waste or an increase in production costs.

Consider the production of paper. The thickness of the paper is an essential criterion in determining its quality. However, because of the high degree of process variability, there are occasions when the thickness quality is lower than the lower specification limit established by the buyers, which results in a drop in sales. High process variability in the production process results in production that is out of limit in which the yields are considered a waste.

Ultimately, the most effective tactic is to find ways to regulate the variability of the production process. A decrease in the variability will result in a lower number of cases that are outside of the limit.

This, as a result, prompts organizations to narrow down the production target range, lowering the cost of the materials, reducing the amount of waste produced, and boosting the quality of products.

Quality Management Tools

Kaoru Ishikawa, who was significant in inventing overall quality management and increasing efficiency in the manufacturing industry by enhancing the quality of deliveries, placed a significant deal of emphasis on the quality management tools listed below (Figure 7.9).

FIGURE 7.9 Total quality management tools and an open team environment are a road to continuous improvement.

dashadima/Shutterstock.com.

These seven rudimentary techniques for quality management are also known as the "original" seven or the "old" seven (Figure 7.10). They are as follows:

FIGURE 7.10 TQM tools are part of manufacturing improvement.

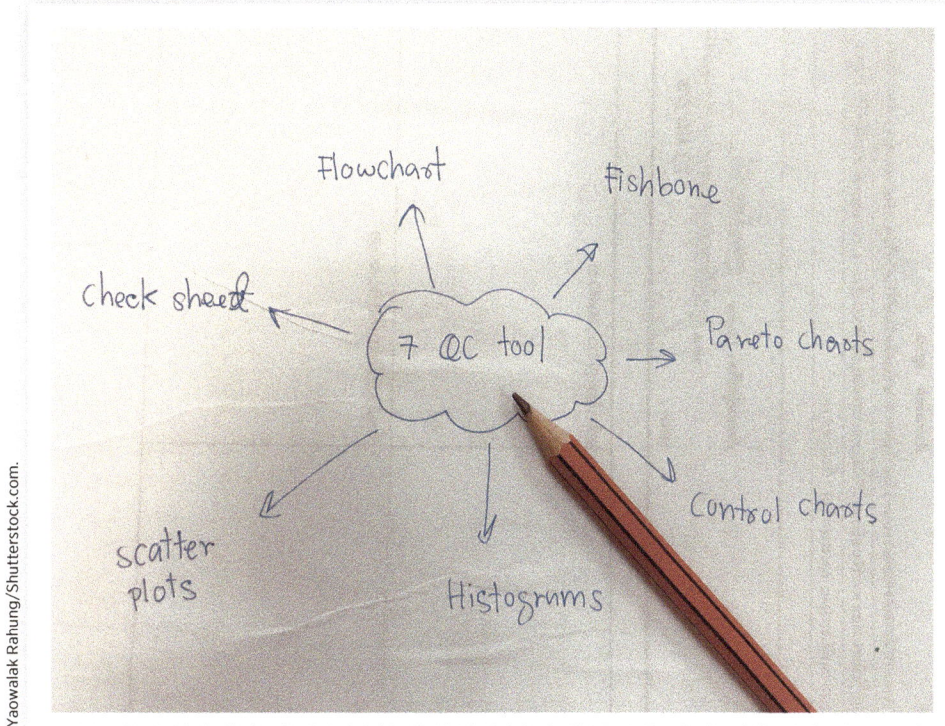

Yaowalak Rahung/Shutterstock.com.

1. **Pareto chart:** A bar graph whose signature attribute is data organized in descending order with a line graph of the summing percent contribution of each. A graphical tool that assists in the dissection of a complex issue into its parts and determining which components are the most vital (Figure 7.11).

FIGURE 7.11 Example of a Pareto diagram.

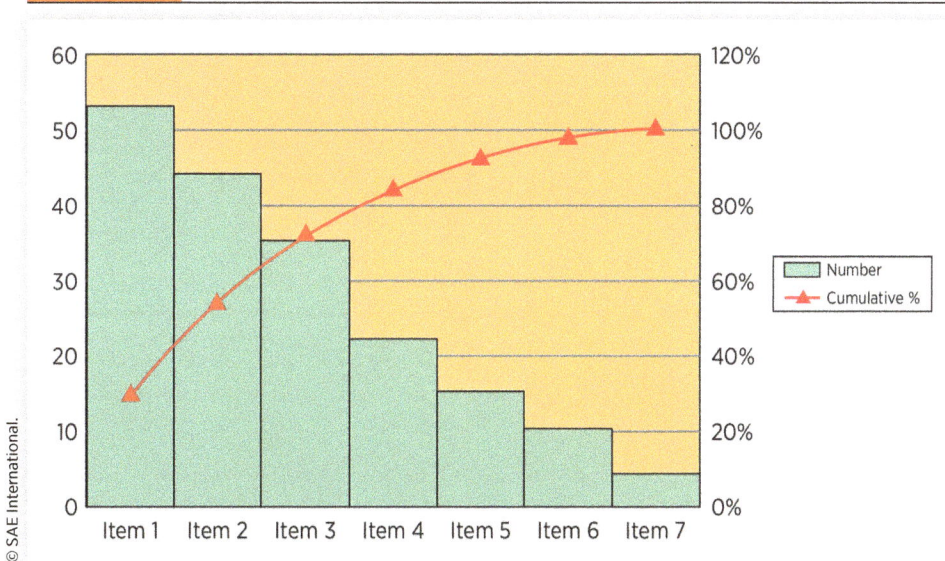

© SAE International.

2. **Scatter diagram:** It is a plot of points used to examine and determine the relationship between two variables, features, or factors. Regression analysis can improve the accuracy of a scatter diagram.

3. **Histogram:** The graph that is utilized most frequently to illustrate frequency distributions, also known as the frequency with which each distinct value in a collection of data occurs (Figure 7.12).

FIGURE 7.12 Example of a histogram.

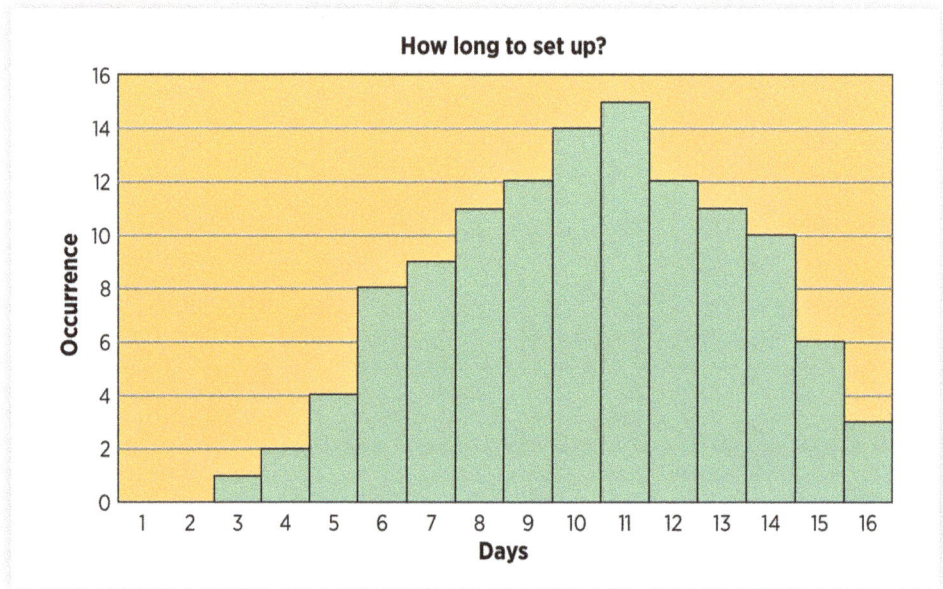

© SAE International.

4. **Cause-and-effect diagram or the Ishikawa diagram:** The problem is described, and then the various causes of the problem, as well as the effect or result of the problem, are listed in the cause-and-effect diagram (Figure 7.13).

FIGURE 7.13 Example of an Ishikawa diagram.

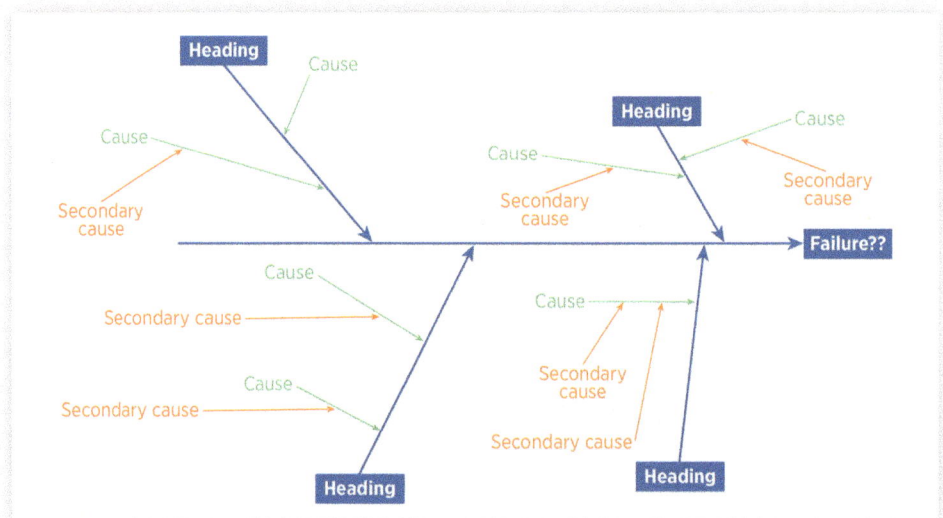

© SAE International.

5. **Check sheet:** It is a document that is used to gather information/data. Even though a check sheet is merely useable for assembling and recording data, the acquired data can serve as a basis for further analysis. Checklists are simple to comprehend and facilitate the transformation of beliefs into facts. Many check sheets are based on the data type and intended application. Each check sheet is created specifically for its intended use. A carefully crafted check sheet facilitates the examination of facts from multiple perspectives.

6. **Stratification:** Stratification is a way to group together many things that may affect the quality of delivery. All the collected data are separated into groups so that different patterns of quality-affecting factors can be made and studied. The stratification method is used a lot when analyzing data to ensure quality.

7. **Control chart:** Control charts are extremely effective monitoring, controlling, and enhancing tools for processes across time (Figure 7.14). They are among the most intricate quality instruments. Control charts are useful for analyzing repeatable operations whose outcomes are anticipated to remain constant over time. Control charts are the fundamental instruments of statistical process control, which has been and continues to be widely utilized in the industrial industry.

FIGURE 7.14   Example of a control chart.

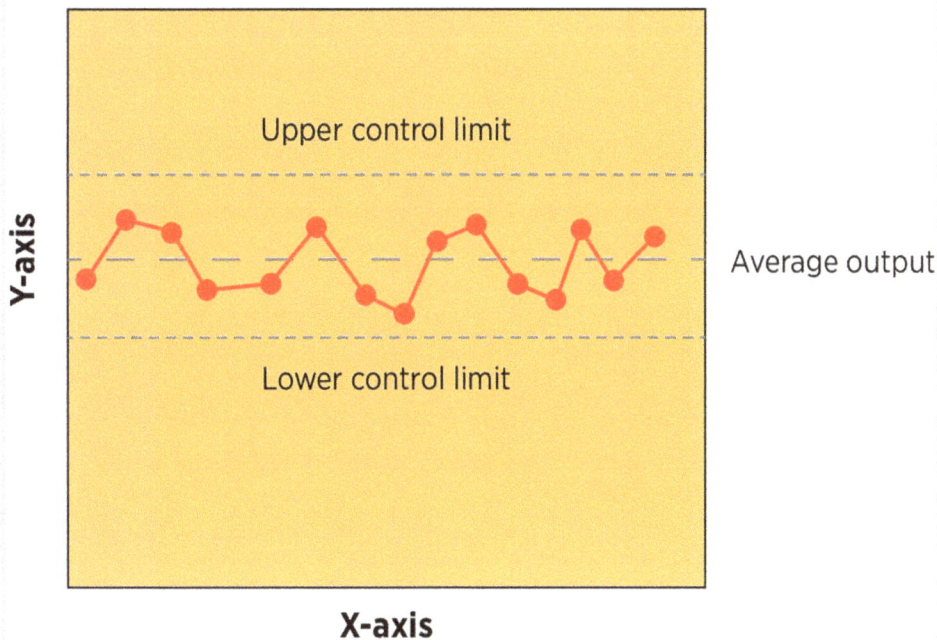

References

1. Baldwin, C. and Clark, K., "Managing in an Age of Modularity," in Gilmore, J. and Pine, J. II (Eds), *Markets of One–Creating Customer-Unique Value through Mass Customization* (Boston, MA: Harvard Business Review, 1997).

2. Baldwin, C.Y. and Clark, K.B., "The Value and Costs of Modularity," in Modularity, Paper of the Opening Conference of Research Institute of Economy, Trade and Industry, IAI (7/2001), June 2001, 5-45.

3. Sanderson, S. and Uzumeri, M., "Managing Product Families: The Case of the Sony Walkman," *Research Policy* 24 (1995): 761-782.

4. Henderson, R.M. and Clark, K.B., "Architectural Innovation: The Reconfiguration of Existing product Technologies and the Failure of Established Firms," *Administrative Science Quarterly* 35, no. 1 (1990): 9-30.

5. Ethiraj, S.K. and Levinthal, D., "Modularity and Innovation in Complex Systems," *Management Science* 50, no. 2 (2004): 159-173.

6. Sanchez, R., "Using Modularity to Manage the Interactions of Technical and Industrial Design," *Design Management Journal* 2, no. 1 (2010): 8-19.

7. Sauer, A., Thielmann, A., and Isenmann, R., "Modularity in Roadmapping—Integrated Foresight of Technologies, Products, Applications, Markets and Society: The Case of 'Lithium Ion Battery LIB 2015'," *Technological Forecasting and Social Change* 125(C) (2017): 321-333.

8. Saaty, T.L., "Fundamentals of the Analytic Network Process," in *Proceedings of the Fifth International Symposium on the Analytic Hierarchy Process, ISAHP*, Kobe, Japan, August 12-14, 1999.

9. Otto, K.N. and Wood, K., "Product Evolution: A Reverse Engineering and Redesign Methodology," *Research in Engineering Design* 10, no. 4 (1998): 226.

10. Kumpe, T. and Bolwijn, P.T., "Manufacturing: The New Case for Vertical Integration," *Harvard Business Review* 66, no. 2 (1988): 75-81.

11. Boothroyd, G. and Dewhurst, P., "Product Design for Manufacture and Assembly," *Manufacturing Engineering* 100, no. 4 (1988): 42-46.

12. Livingston, I., "Design for Service," in *Proceedings of the First International Conference on After-Sales Success*, London, UK, November 29–30, 1988, 45-71, ISBN:1-85423-0289.

13. Kossiakoff, A. and Sweet, W.N., *Systems Engineering: Principles and Practice* (Hoboken, NJ: Wiley, 2003).

14. Suzik, H.A., "GM Announces Modularity Project," *Quality* 38, no. 5 (1999): 14.

15. Kusiak, A., "Integrated Product and Process Design: A Modularity Perspective," *Journal of Engineering Design* 13, no. 3 (2002): 223-231.

16. Fisher, M., Ramdas, K., and Ulrich, K., "Component Sharing in the Management of Product Variety: A Study of Automotive Braking Systems," *Management Science* 45, no. 3 (1999): 297-315.

17. Feitzinger, E. and Lee, H.L., "Mass Customization at Hewlett-Packard: The Power of Postponement," *Harvard Business Review* 75, no. 1 (1997): 116-121.

18. Onkvisit, S. and Shaw, J.J., "Service Marketing: Image, Branding, and Competition," *Business Horizons* 32, no. 1 (1989): 13-18.

8

System of Systems

Systems Engineering

This chapter is about systems engineering (SE) and product and manufacturing line design exploration (Figure 8.1). SE facilitates product design and manufacturing line development exploration and growth, including documentation traceability to subsequent testing (via configuration management [CM]). Rather than focus on the product or manufacturing design, SE focuses on how the product fits in the entire context.

FIGURE 8.1 Vehicles are a collection of systems.

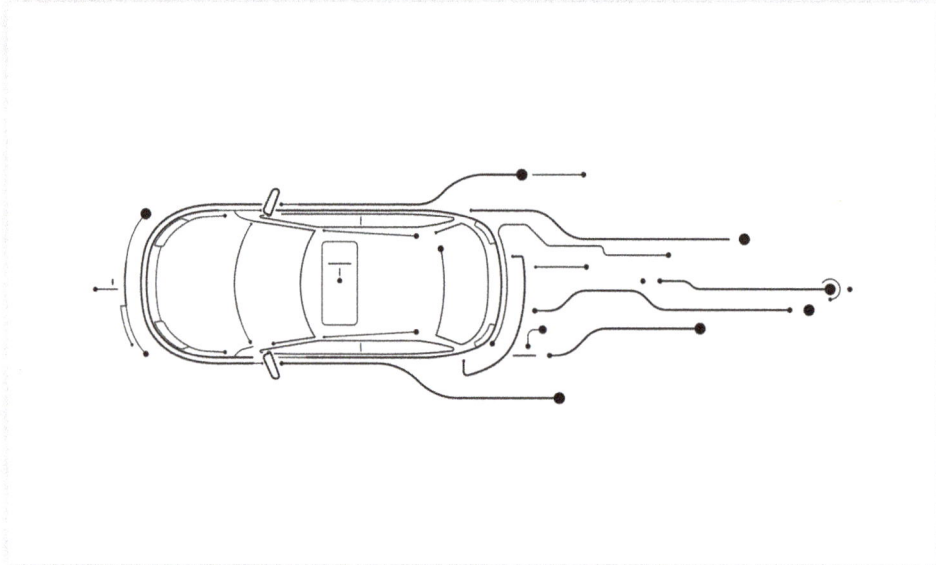

The testing will be increasingly more complex and broader in scope. From the Department of Defense (DOD) Systems Engineering Guidebook, we have the graphic Figure 8.2:

FIGURE 8.2 Systems engineering from DOD.

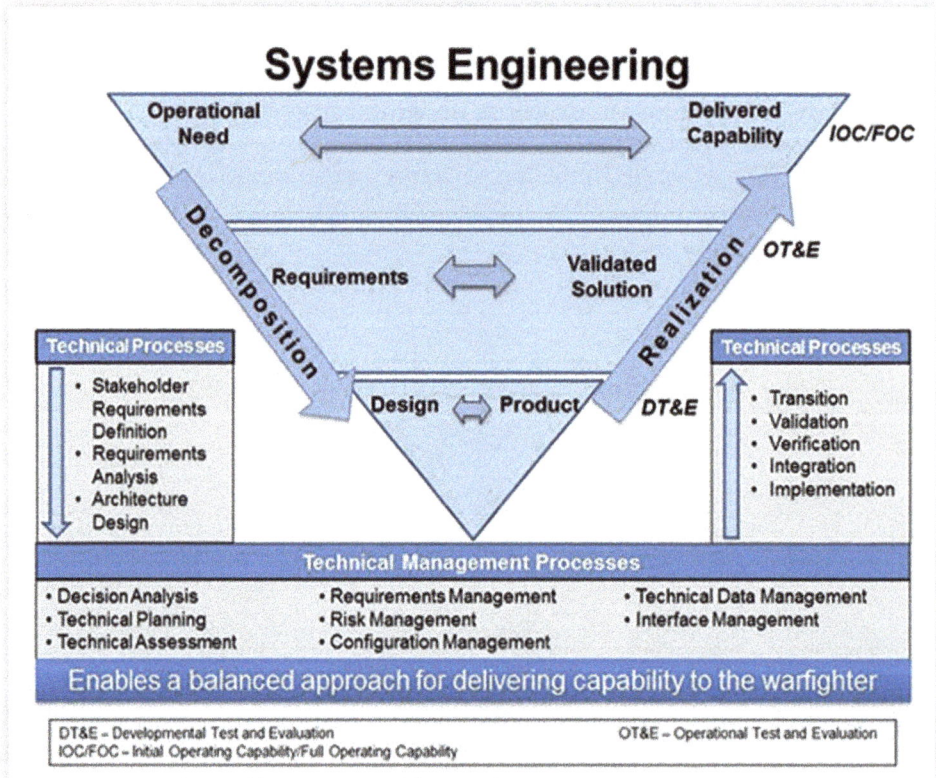

What Is Systems Engineering?

Systems engineers tend to serve as technical oversite of everything that occurs on a project, and it may look like they add too much process overhead and non-value-added work. A senior systems engineer from a significant US corporation went to each division to promote the use of practical SE techniques. His message included a description of what SE can and should do to commercialize products. He also strongly emphasized planning and documentation in his speech. Over several months, he met with division managers, chief engineers, program managers, and senior engineers. He came back absolutely without enthusiasm. The idea was completely ignored; the proposal appeared pointless work or was well beyond what they believed they could reasonably afford to spend their time and money. A while later, a different senior systems engineer paid many of the same people a visit for the same reason. This engineer's message was that significant improvements might be obtained by concentrating on the most crucial customer needs and using a small number of SE tools/practices. The message was well received this time [1].

Tracking back to the DOD Systems Engineering Guidebook (Figure 8.3), we see an overview of the stages of development as well as reviews and other checkpoints to ensure product quality.

FIGURE 8.3 DOD SE overview.

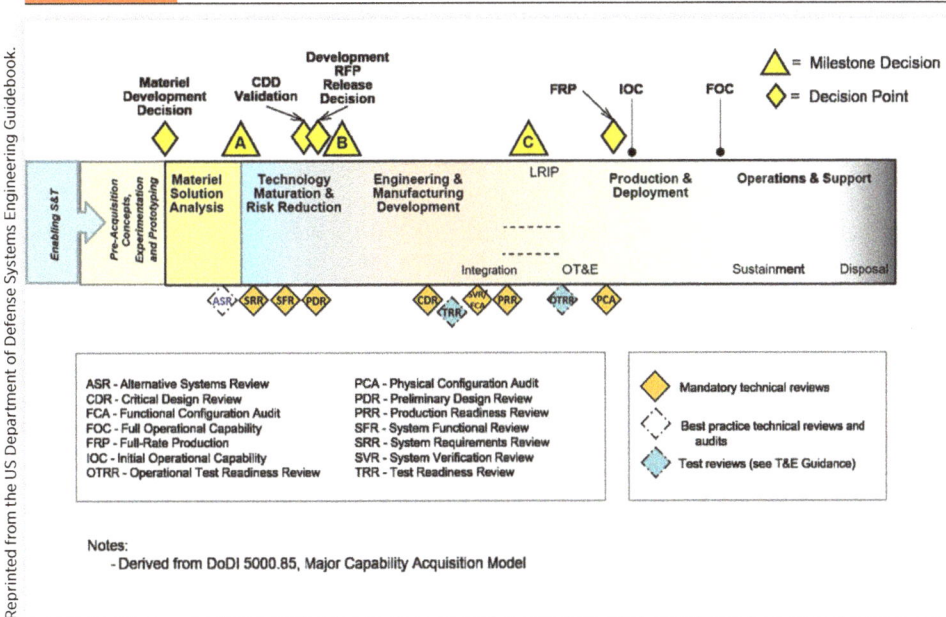

Notes:
- Derived from DoDI 5000.85, Major Capability Acquisition Model

The Value of the Early Whiteboard Exploration

The ability to estimate the effort and expense of software systems is aided by early and precise program size estimation. The function point analysis (FPA) method is often employed for estimating program size. Software complexity, effort, and cost associated with software development are all usefully measured using software size and effort estimation method-ologies. The research community has been studying software estimation for almost three decades, but it has not been able to produce a trustworthy estimation model for end-user development (EUD) contexts. In essence, EUD outsources development work to the client. Because of this, the additional design time required for end-user programming is one factor in the size and effort. However, the problem is not just for software projects but any project where we do not understand the technology or the specific application of the technology.

It is extremely common to size software according to its function, as in function points (FPs). However, it is worth noting that FPA was not created because a new system size metric was needed; instead, it was developed because productivity was essential in today's society [2].

What Can Go Wrong?

Many expensive initiatives have been abandoned because of the complexity of massive engineering projects, and implementations have been severely hampered in other situations. The complexity of the project itself leads to these failures, for example, team distribution. Complex system development demands an evolutionary method involving people and technology (hardware and software) as active participants in the system evolution. Rapid adjustments must be possible during this evolutionary process while maintaining robust-ness and safety. The study of complex systems offers two solutions to the failures of substan-tial engineering projects. The first step is to alter your goals.

Given that this complexity is often an intrinsic characteristic of engineering challenges, planners should strive to keep the complexity of their aims to a minimum. The structuring of successful projects depends on this. Using an evolutionary process is the second option. An alternative approach becomes crucial when simplification is no longer practical because of the underlying complexity of the needed function, which implies significant degrees of function uncertainty within the bounds of rationality and modeling. In this situation, various potential methods can be explored methodically, enabling the development of highly complex creatures [3].

It is not just about what can go wrong with the process but also the product or how it fits into the system. For example, one of us wrote systems specifications for advanced braking systems such as adaptive cruise control or hill hold for a heavy commercial vehicle. When exploring how the system would respond given a sensor failure, we discovered that in the system as articulated. This sensor failure would drastically and negatively affect the perfor-mance of the system.

Now we have a paper design problem; the proposed design is in the paper phase, and we know there to be a system design issue requiring attention (Figure 8.4).

FIGURE 8.4   How do we want things to fail? What is the response?

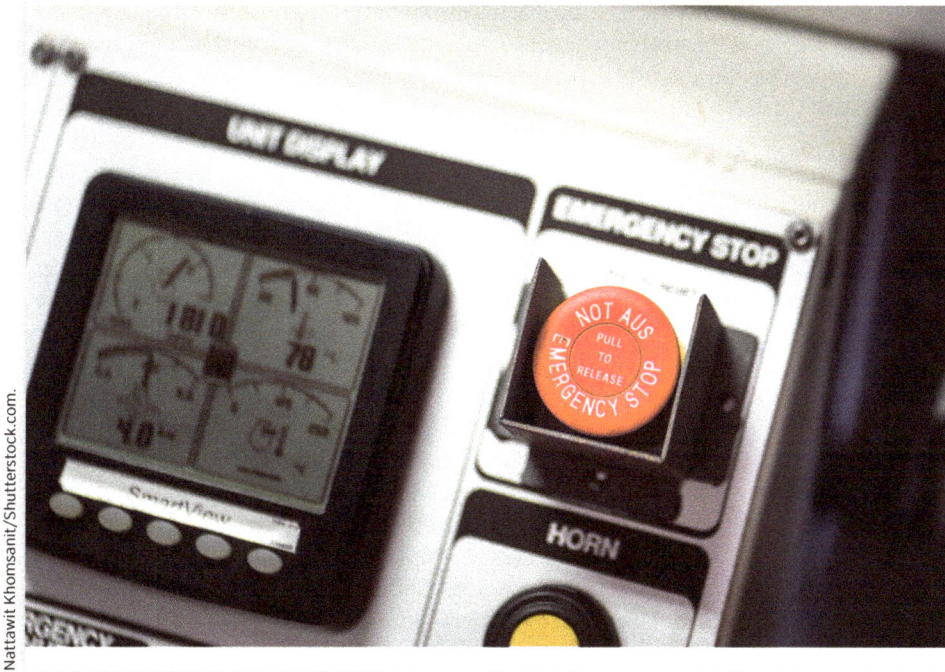

Nattawit Khomsanit/Shutterstock.com.

What Response Do We Want When Things Go Wrong?

One of the most popular methods for determining the size of projects or software systems is FPA. Each function is categorized according to its relative functional complexity throughout the point-counting procedure, indicating the dimension of a project or application. Research has already suggested extending FPA to improve point assessment precision for systems with more complicated algorithms. This article means developing FPA to fuzzy function point analysis (FFPA) using ideas and characteristics from fuzzy set theory [4]. The goal of fuzzy theory is to provide a formal, quantitative structure that can mimic the uncertainty inherent in human understanding. As a result, derived quantities like costs and terms of development can be more precisely calculated using the FPs produced by FFPA.

The management of projects is vital to software engineering. It includes the complete software development process, clarifies and organizes it, and coordinates staff activity in executing processes methodically and effectively. The project manager must specify things like the length of time for development and the price of the finished product once the scope of the project has been established. These actions are crucial because if these estimates are too high, the project may not be accepted when it perhaps should be. The development process may experience emphasis changes that compromise quality or diminish the functionality of the finished product.

Software projects frequently experience worst-case scenarios marked by ballooning deadlines and prices. One of the best practices for software metrics starting in 2001 is FPA. Since then, FPA has acquired an expressive application in software project management, especially as a result of its technological independence, simplicity, and conciseness. A project or application is broken down into its data and transactional functions as part of the FPA process. While the transactional functions define the functionality supplied to the user with respect to the application processing of this data, the data functions reflect the functionality provided to the user by adhering to their internal and external requirements concerning the data.

Each function must first be identified before being categorized as either simple, average, or complex based on its relative functional complexity, which is indicated by a specific value in points, depending on the function. The application is adjusted in accordance with the general features of the system once all functions have been scored, which assesses the overall functionality of the application. FPA classifies its functions in a sudden and haphazard way as initially intended. For instance, an external input (EI) with two file references and five data items is rated as "average" and awarded four points. With four points, another EI that makes use of two files and fifteen data points is rated as "average." In another instance, the function is deemed "difficult," earning six points, when it refers to two files and sixteen data elements. Two severe issues are instantly apparent in such a categorization scheme. First, functions of different sizes obtain the same point values, and second, similar functions are suddenly divided into distinct groups. When many system functions fall within the boundary regions of the given intervals, this inconsistency may get even worse.

Governmental systems, for instance, particularly those that manage tax payments or check the sincerity of legal entities in their financial dealings, typically have a lot of transactional functions that cross the line into high complexity by referencing a lot of files and data items in a single basic process. This fact unites functions that "intuitively" have distinct widths. Because of this, the FPA estimates suggest that these functions should be included in identical developmental timeframes and costs, which is not supported in practice. The development of software is another important issue. No matter how large, it is impossible to create a system without considering the need for adjustments in the future. The original requirements for software are altered during the programme lifecycle to account for user changes and client needs. Statistics show that many companies devote at least 50% of their financial resources to software maintenance since the end of the 1980s. This demonstrates the significance of research into software maintenance. To modify the activities required for this particular process, new methodologies are developed across them or existing methodologies are modified.

Many studies have already advocated expanding FPA, such as FFP, with the main goal of achieving more accuracy in the point evaluation of systems with more complicated algorithms, such as real-time systems, embedded systems, and communications systems, among others. Some extensions are exclusive to improvement initiatives. It was suggested that the FPA structure be changed to allow for a more accurate assessment of tiny functional increments. Contrary to traditional estimations, there seems to be growing interest in investigating methods that could replace or supplement the ones already in use. These methods use intelligent resources such as neural networks, case-based reasoning, regression trees, analogies, and fuzzy logic. To convert FPA into FFPA, it is suggested in this study that notions and properties from fuzzy set theory be used (FFPA). Building a formal quantitative structure that can capture the imprecision of human knowledge or how knowledge is communicated in natural language is the main driving force behind the fuzzy set theory. This theory aims to close the gap between the conventional mathematical models of physical systems and their frequently inaccurate mental representations [4].

Responses to Errors

Allocating attention and resources to the components and subsystems that are the most unreliable and prone to failure is one of the fundamental challenges in maintenance. The fault tree analysis (FTA) technique can be used in industrial systems to examine the dependability of complex systems and their substructures. This study presents a fault tree application that may be used to analyze the existing reliability and failure probability in real time for maintenance needs [5]. Data related to the fault tree root causes and events are used in the analysis. A new determination of the current probability of the fault tree events and subsystem interactions can be brought about by an indication of an anomaly case, service action, cumulative loading, etc., or simply by time passed or service hour counter level. Real-time information from various data sources is coupled to each fault tree event and root cause in the proposed technique using dynamic information.

To help maintenance decision-making and to keep the analysis current, a live, constantly updated interface between fault tree events and maintenance databases has also been developed. Usually, updated root cause probabilities and lower-level events are used to evaluate top event likelihood. At the industrial plant level, the failure probability of an event described within a created and operational fault tree is explicitly indicated by the identification of a failure in that event. The most likely failure branches through the fault tree subevents to root causes can be determined using this signal, and the fault tree branches that are most likely to be the cause of the failure can therefore be targeted for service first. When components, particularly those found in the critical branches, are discovered as healthy by inspections, service activities, etc., they can be modified to have zero failure probability [5].

Detection

Detection Modes

1. Post-processing
   - Visual inspections
   - Experience
   - Monitoring and sampling
2. Testing (QA)
   - Static
   - Dynamic
   - Formal or regression testing (test loops)
3. Comparison of analytical results (if available)
   - Experimental wind/water tunnel testing
   - Previously published and peer-reviewed studies
   - Benchmark
   - Google

Prevention

Defect prevention is the most important but least prioritized aspect of every software quality assurance of a project. If used throughout the entire software development process, it can cut down on the time, money, and resources needed to create a high-quality product. The most successful and efficient method for fault detection and prevention is software inspection. Inspections conducted across the whole software lifecycle have proven to be quite effective and add value to the product qualities. If the clients are completely satisfied with every transaction, the IT industry will be successful. If the company can manufacture a high-quality product, this is feasible. A product must be defect free and deliver the desired results to be considered high quality. It should be delivered for the estimated price and time and manageable with few interruptions.

Typically, a minor increase in preventive measures will result in a significant reduction in the overall cost of quality. However, the primary goal of a quality cost analysis is not to lower costs but to ensure that the costs incurred are the appropriate ones and to maximize the benefits obtained from the investment. Defect prevention has become the main focus as a result of quality cost analysis. Additionally, it has been seen that as defect detection and prevention practices are used, over time and at a certain optimum point, company performance improves, quality rises, and the cost of quality falls. Any flaw or imperfection in a software work product or software process is referred to as a fault. A defect is an oversight, flaw, or failure. The IEEE/Standard defines the following terms as Error: human mistakes that produce inaccurate outcomes. Fault: making the wrong choice when interpreting the information provided, trying to address issues, or carrying out a process. Failure is when a function is unable to fulfil the requirements as intended.

Need for Defect Prevention

Early defect analysis cuts down on the time, money, and resources needed. The ability to prevent defects is made possible by an understanding of defect injection procedures and methods. When this information is properly applied, the quality rises. Additionally, it raises productivity overall.

Advantages of Preventing Defects

All IT businesses must strive to cut down on errors as much as possible. The presence of fault prevention techniques indicates a highly developed test procedure. The prevention of errors moving from requirement specifications to designs and from designs to code is made possible by the early detection of faults in the development lifecycle. Defect prevention provides significant time and money savings during the application development process. As a result, it dramatically lowers the number of defects, reduces the cost of rework, makes system maintenance and reuse simpler, increases system reliability, and requires less time and resources for the organization to produce high-quality systems. Based on the lifecycle stages in which defects were injected, preventive measures are discovered, and productivity is also increased [6].

Fault Tree Analysis

The method used for the reliability analysis of complicated systems is FTA. The basic idea is to translate a model failure behavior into a visual diagram or logic model. The most popular method for causal analysis in risk and reliability research is FTA. To statistically assess the likelihood of a safety hazard, this analysis technique is mostly applied in the field of SE. To determine the likelihood of the top event, a logical approach is used. FTA, or failure analysis, is the process of employing Boolean logic to analyze an undesirable state. H.A. Watson created FTA in 1962 at Bell Laboratories as part of a contract with the US Air Force Ballistics Systems Division to test the Minuteman I InterContinental Ballistic Missile (ICBM) Launch Control System. The dependability specialists broadened the FTA concept. From 1963 to 1964, Boeing and Average Cost Method (AVCO) expanded their use to include the Minuteman II system. Since then, it has been employed in a variety of fields related to system safety assessment and reliability engineering, including nuclear reactors, the chemical and manufacturing sectors, circuit boards, and the petrochemical industry.

FTA involves deterministic contributions to events resulting from assigning failure rates to the branches, such as events caused by material failures, hardware wear out, or combinations of these. Failure rates are calculated using the component, unit, or subsystem Mean Time Between Failures (MTBF). FTA can be used as a design tool to spot accidents and as a diagnostic tool to foretell system failures most likely to occur. The fault tree offers a diagrammatic depiction of the potential failure modes of a system. The fault tree is crucial for safety system analysis since it produces an exhaustive list of all possible causes of system failure. As a result, the engineers can spot and fix any design flaws [7].

Fault tree has continued to be one of the most frequently utilized fault management (FM) techniques by practitioners throughout the years (Figure 8.5). It is a powerful visualization/communication medium and a quantitative analysis tool for developing reliable systems. However, because of potential misrepresentation of the links among failure events, FTA has frequently been unable to give us high-confidence conclusions. In hindsight, combined fault manifestation and interaction almost invariably resulted in catastrophic system breakdowns. Additionally, flaws in FM itself may play a crucial role in a chain of fault manifestation. We can learn from the tragic results of improper FM implementation from the 1996 self-destruction of the Ariane-5 rocket. We have proposed a fault-class-aware and FM-capability-aware FTA paradigm as a result of these preceding lessons [8].

FIGURE 8.5   Example of a fault tree diagram.

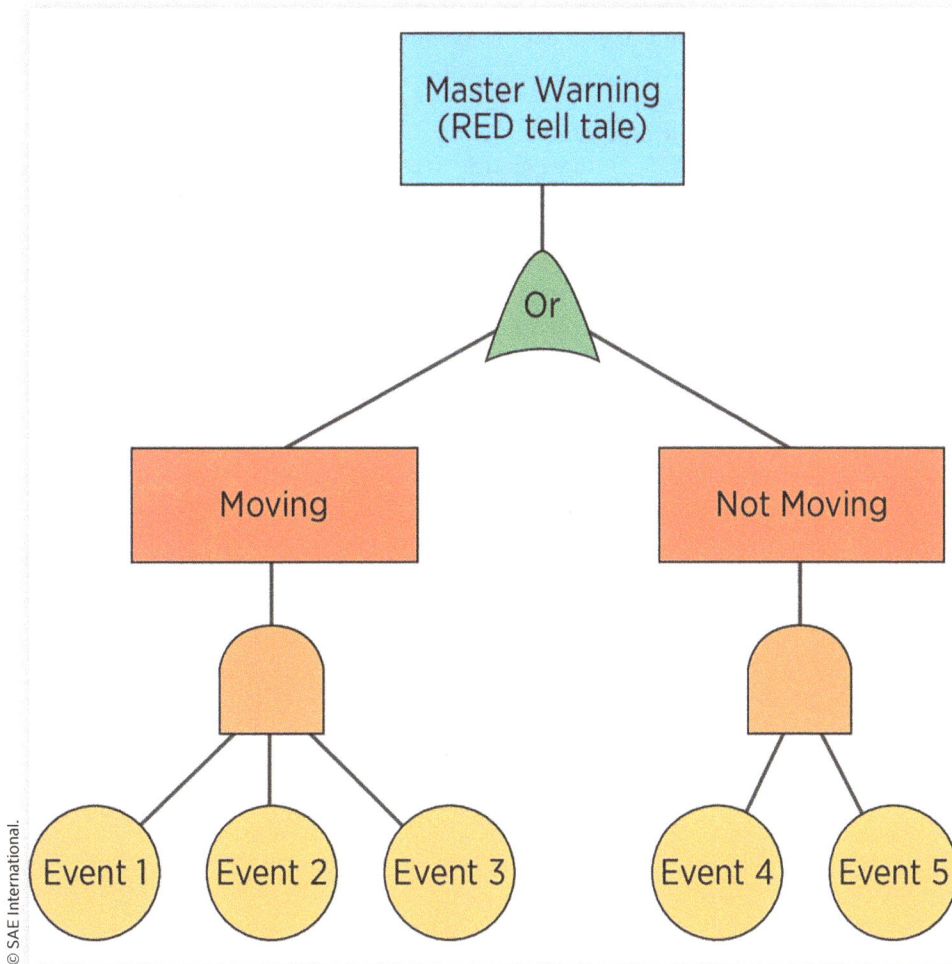

Function Point Analysis

What Is It?

A software size can be calculated using the FPA standard measurement technique. The FPA results are consulted for estimating project costs or effort, benchmarking, contract outsourcing, progress tracking, estimating portfolio size, negotiating change, and managing defect density. The FPA regulatory body, the International Function Point Users Group (IFPUG), is in charge of the Counting Practices Manual (CPM). ISO/IEC 20926:2010 has standardized the FPA1 of the IFPUG. There have been several reported applications of the FPA. Concerns about the advantages and disadvantages of FPA are in dispute. FPA is heavily criticized since it relies on individual judgment, which prevents a standardized application. Differences of 30% were observed for the same product, whereas variations of 12% were reported for counters within the same organization. Even more, variation is noted among organizations. Others contend that, in addition to the fact that APF can be expensive and time consuming, subjectivity is a significant issue.

Most solutions to the FPA subjectivity aim to minimize human involvement. To determine the functional size, they provide a measurement process that uses mapping rules between artifacts developed using software modeling techniques and FPA ideas. These methods help to increase measurement repeatability across a range of measured quantity values; however, they oversimplify the IFPUG's CPM criteria. This simplification is necessary since none of the used artifacts are sufficiently detailed to allow for the full use of the conventional FPA approach. Since these models were not created with FPA in mind, the accuracy and completeness of existing artifact models, like Unified Modeling Language (UML), are not guaranteed. The necessary information might not be provided by models that were not expressly designed to allow measurement.

In the worst circumstances, the artifact model lacks a crucial piece of information that the rules demand, making the FPA stages inapplicable. The measurement accuracy in relation to the genuine quantity value, which is defined as the anticipated number of FPs by the appropriate FPA application, is thus compromised by the present ways of improving reproducibility. When compared to the actual quantity value, reproducibility and accuracy refer to confirming the consistency and concordance of measurement findings obtained from repeated measurements taken by various people under similar or the same conditions [9].

How to Do It

The following steps are part of the application process for FPA, according to the IFPUG. (1) Specify the boundaries, scope, and objective of the counting: A logical boundary between the software being measured, its users, and other software systems defines a software boundary. It is based solely on the user's external business perspective since it is the only factor that matters. It might be challenging to distinguish between one software and another because the boundary can be subjective. Therefore, the border should take a business viewpoint rather than only technical ones. (2) Measure data functions: A data function takes care of a user's functional needs for referring or storing data. Internal logical file (ILF) and external interface file (EIF) are two different sorts of Business Function Coordination (BFC) that can be categorized into. It is necessary to group data tables from the logical side, classify them as ILF or EIF, count the number of data element types (DETs) and record element types (RETs) in each ILF and EIF, and then determine the complexity (high, medium, and low) and functional size of each BFC data function to identify the existing data functions on the software boundary. (3) Measure transactional functions: Software capability made available to the user for data processing is referred to as a transactional function. There are three sorts of BFC that falls under external input (EI), external output (EO), and external inquiry (EQ). The steps to measuring a transactional function are as follows: (a) break down the requirements into the smallest units of activity; (b) classify each transactional function as EI, EO, or EQ; (c) count the file type references and the number of DET for each EI, EO, and EQ; and (d) calculate the complexity (high, medium, or low) and functional size of the transactional function for each BFC. (4) Determine the functional size: The functional size can be determined using a variety of formulas depending on the project type, scope, and objective (development, improvement, or application). (5) Record the counting and present the findings: The tracking of the functional size obtained from counting using FPA is ensured by recording the premises and interpretations during the counting. This strengthens the method consistency and highlights potential areas for improvement in the counting documentation [10].

Areas of Improvement

Software companies are working to increase the project cost estimation accuracy. In turn, this aids in resource allocation. The software engineering community has long been interested in software cost estimation. Over time, many estimation models in a number of categories have been put forth. FP, a valuable tool for estimating the cost of software projects, was initially put forth 25 years ago utilizing a project repository that contained data on many facets of software projects. Although the productivity of software development has increased significantly over the past 25 years, the complexity weight metrics values given to count standard FP have remained constant. The validity of the complexity weight values and the precision of the estimation method are seriously questioned by this fact. With the use of the project repository of the International Software Benchmarking Standards Group dataset, this work aims to propose a genetic algorithm–based method for calibrating the complexity weight metrics of FP. The contribution of this work demonstrates that the software estimation process is more accurate when information is reused and function-point structural aspects from previous projects are integrated (Figure 8.6).

FIGURE 8.6   SE facilitates continuous improvement of products and processes.

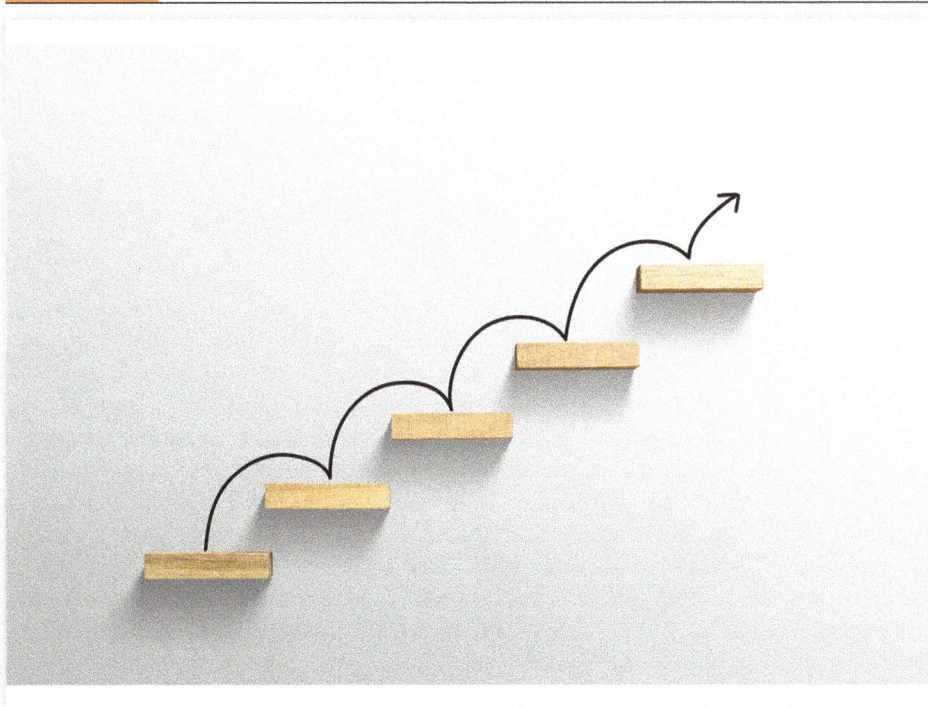

Shutter_M/Shutterstock.com.

A variety of factors influence the success of software initiatives. One of the elements is early development phase effort estimation. Since resource allocation is so highly dependent on this activity, accurate software estimation is essential for project success. On the other hand, an incorrect estimate poses a risk to the success of the project and a risk to the livelihood of even a tiny software company. One may argue that Constructive Cost Model (COCOMO), Software Lifecycle Management (SLIM), and FP are three of the most well-known models for estimating software development efforts that the software engineering research community has presented. Algorithmic models are what these models belong to. This is so that the algorithms used to calculate development effort estimates can be calibrated using previous information acquired from the software industry. Because it can be used in the early stages of development, such as requirements engineering and system analysis, and because it is programming language independent, FP is a great way to estimate the cost of the software. The functional size of the software is reflected in the complexity weight metrics values of FP. Based on historical data from software projects, they were first introduced in 1979 and are now used globally. Software development techniques have gradually advanced throughout time.

Today's software development approaches vastly differ from those used more than 20 years ago. As an illustration, the object-oriented paradigm has become a standard in software development. The complexity of creating programs with graphical user interfaces has decreased because of visual programming languages. Database applications are user-friendly because automated report production technologies make it simple to create complex reports. The amount of time spent on the software development lifecycle has decreased even further thanks to Computer-Aided Software Engineering (CASE) tools. The necessity of calibrating the complexity weight metrics of FP is further supported by the fundamental changes in software development since the complexity weight metrics of FP were established 25 years ago. Integrating software project data from the recent past will make this calibration more thorough and precise [11].

Automating FPA—Model Development

A common software measurement called FPs is used to quantify the features that a program provides to the user. Allan Albrecht developed FPA for IBM, which has since replaced popular structural size measurements (like LOC) as the method of choice for measuring functional software. FPA is more effective in terms of structural measures because of its close ties to development expenses and the potential for an early evaluation of the production effort. FPA has developed over the past 20 years into a more specific and defined standard, and many amendments to the original definition put out by the IFPUG have been suggested. The most widely used interpretation of Albrecht's suggestion continues to be the most recent iteration of the IFPUG standard, which is found in the so-called Counting Practices Manuals.

The IFPUG counting method is based on categorizing the functions that the system is expected to carry out and putting a corresponding complexity level on each one. The many functions are divided into five groups according to their main purpose. Data functions—functions that deal with stored data—are further divided into ILF and EIF, depending on whether the associated data are handled internally by the system or externally by the application. Transaction functions are those that involve interacting with other users or external agents. They can be classified as EIs if their main goal is to collect information from users, as External Queries if their goal is to retrieve information that has been requested by users, or as EOs if they provide users with information that is the outcome of an action or elaboration. The system's identified functions are then examined to see what its fundamental components are. It is required to assess both the RETs, which are collections of data items of particular importance to the user, and the DETs, which are the smallest data pieces meaningful to the user, for each Data Function [12].

Reproducibility

FPA is a standardized method for determining software size based on functional needs, considering functionality that will be created in response to user requests and responses. The results of FPA are frequently used as a benchmark for calculating other quantitative factors, such as effort, productivity, or cost. The FPA regulatory body, the IFPUG, is in charge of enhancing and developing the guidelines outlined in the Counting Practices Manual CPM (IFPUG 2010) [13], version 4.3.1. Additionally, FPA is codified in ISO/IEC 20926:2010. However, a common complaint is that FPA is too subjective because it relies on expert judgment, which limits its application to standards, as covered in related work. Most solutions to this subjectivity include applying rules to translate between FPA concepts and software modeling artifacts (like UML) [10].

Existing artifact models were not created with FPA in mind; therefore, there is no guarantee that they are accurate or comprehensive. These methods oversimplify the CPM principles even if they at least somewhat improve measurement repeatability between various metrics. This simplification is necessary since no artifact is sufficiently detailed to use the IFPUG's FPA as the standard FPA method. In the worst circumstances, the artifact model lacks certain crucial data, making the FPA counting rules inapplicable. The measurement accuracy in relation to the genuine amount value is thus compromised by the current ways of improving reproducibility. Reproducibility and accuracy are closely linked concepts that pertain to confirming the consistency and concordance of measurement findings received from repeated measurements, by various people, under comparable or identical settings [9].

Integrated Development

As a result of increased coordination requirements brought on by integrated product development (IPD), other aspects of the NPD process (such as integrated tools), product definitions (such as incremental development), organizational context (such as reduced task specialization), and teams are used to make up for these shortcomings (e.g., cross-functional teams). IPD is now a crucial new standard for overseeing the development of new products.

One of the most important modern developments in the management of NPD is IPD. IPD capabilities, such as increasing customer involvement and developing more producible designs, should boost one of the most important modern developments in the management of NPD, that is, IPD. In actuality, IPD has evolved into the model for NPD. Griffin [14], for instance, discovered that cross-functional teams, one of IPD's key qualities, were utilized in 64% of the projects she examined. The performance advantages of the new paradigm over conventional approaches are probably what are driving its expansion. Our third goal is to offer potential future research themes that should entice researchers to move into significant new contexts. The IPD features, such as increasing customer involvement and producing more manufacturable designs, should bring projects in a single company. IPD researchers should find benefits in achieving all three goals. Companies interested in or using IPD would benefit from achieving at least the second goal [11].

Compare and Contrast

Why should a company be concerned about processes? Processes and measuring are the mechanisms by which we comprehend, assess, manage, learn, communicate, enhance, foresee, and validate completed tasks. How can businesses improve their processes? They have the ability to record, describe, gauge, evaluate, compare, and alter them [1].

INCOSE

Consider "evolving" the system rather than "designing" it (on a clean sheet of paper). Complex systems are modified versions of other complex systems rather than being constructed from the start. The original requirements are probably going to be unclear, and systems-of-systems engineering is never complete, according to INCOSE.

It is incumbent upon systems engineers to be aware of the complex SE practices, and to improve upon those existing SE practices (a). INCOSE may make a difference by adapting to the setting where systems engineers learn to apply complex systems concepts and by purposefully interacting with the complex systems research community (b). The best scenario will be if team members know the guidelines for changing SE practices (c). It means that there should be several available versions of sound SE practices that compete with one another during routine SE practice.

It is time to comprehend and start utilizing ideas from the center and the chaotic side of the spectrum because, up until this point, SE has only focused on using the "order" side of the order-to-chaos spectrum. INCOSE, the SE procedure (such as a company standard procedure), and air traffic control are three examples of complex systems. INCOSE represents most social and voluntary organizations. Most systems engineers are unaware that a company SE process is a network that may be investigated using complex systems methodologies. The definition of a system as it is used by many system engineers is air traffic control [15].

Manufacturing

Traceability of Manufacturing Line Design

Complexity and uncertainty are two of the key manufacturing-related issues. Real occurrences, such as actual demand and output, do not match time-phased plans that have been produced by some ideal model or simplified planning logic because of unpredictability and complexity. The impacts of complexity and uncertainty are attempted to be addressed, and real-world occurrences are attempted to be reconciled with the planned program, using a variety of strategies. These include formalized decision-making, simulation, various degrees of forecasting, performance monitoring, and replanning. The accuracy of the tracing data available for the actual system under consideration determines how effective these techniques will be.

Traceability, in general, is the capacity of a system to reveal either the present or past status of actions. Because it is used in the ISO 9000/BS 5750 quality procedures, the phrase "traceability" has gained popularity recently. Traceability in these situations especially refers to the capacity to trace actions and confirm the occurrence of specific occurrences. The method of traceability proposed in this article takes a much broader perspective that goes well beyond historical record maintenance. We discuss the information needed from the tracing functions at various stages of the event history and in the manufacturing system. The manufacturing system is divided into operations, planning, and strategy. We identify five interconnected management activities, including tracing functions, that are necessary to respond to discrepancies between the anticipated schedule and the events recorded at each level and in each subsystem. We propose three significant types of traceability—status traceability, performance traceability, and goal traceability—after defining the ideal tracing functions of a manufacturing system.

Tracing Function Requirements There is a saying that goes, "You cannot manage what you cannot manage; you cannot control what you cannot measure." In any factory management system, detecting and measuring are two crucial processes. Taking a broader perspective, we can classify these two actions as part of the system tracing function. Therefore, it is vital to examine the system structure and its primary goal, the timely execution of manufacturing events, before discussing the general tracing needs of a manufacturing system [16].

Configuration Management

Overview of Areas

CM Planning Making the enterprise's aims for customer happiness evident is the aim of the quality management process. To ensure that they fulfill quality goals and customer expectations, enterprise policies and procedures control the products, services, and implementations of the system lifecycle (SLC) activities. Every company must have a quality management process because time, money, and quality are the three main factors in any project. A large portion of the motivation for investing time, money, and effort into implementing these processes in the business stems from the fact that many SLC processes are concerned with quality issues. One strategy for introducing a quality discipline into an organization is to use this manual. The emphasis on customer satisfaction and corporate goals and objectives is established, put into practice, and continually improved. Managing quality has a price as well as a profit. The time and effort needed to control quality should not be greater than the total benefits of the process.

Configuration Identification Because of the inherent complexity of the tokamak design, the many systems that are all necessary for its operation, the worldwide distribution of the design activities, and the unusual procurement strategy based on a combination of in-kind and directly funded deliverables, the construction of ITER will present a significant challenge for the fusion community as a whole. A systematic technique to ensuring the consistency of the design with the needed performance must be followed if such a big project is to succeed. To facilitate communication and collaboration among the institutions and industries involved in the project, appropriate project management methodologies, tools, and working practices must be used. The authors were involved in the SE process during the selection and optimization of the machine configuration as well as the definition and actual application of the design integration and configuration control structure inside ITER. Meanwhile, they evaluated design, drawing, and documentation management tools for the building stage.

How to identify the layout of the plant and get the technical data of its components is the first problem that must be solved in a huge project like ITER. When designing mechanical devices, one often creates a small number of top-level assembly drawings that show the entire system (the system configuration) and link to other drawings and production instructions. These drawings also typically include a parts list [17].

Configuration Control The final cost of a large project requiring significant technological effort is primarily decided by (a) the technical choices taken early in the project development and (b) the way adjustments are managed. The combined term "configuration management" or "configuration control" is often used to describe these two actions. Confusion, duplication of work, timetable slippage, and expense overruns are invariably the results of poorly based decisions and wrong or inadequate design criteria. On the other hand, poorly managed adjustments can cause serious issues with budget and schedule. Both requirements must be met for a project to be successful. The correct management of both of these components—the crucial technical choices and the unavoidable change proposals that will surface as a project moves forward—is one of the SE key responsibilities.

Establishing the process or procedure by which those changes will be examined, and implemented, and how choices will be made on what modifications will be accepted and which will be refused is the first stage in establishing effective control over changes that will be necessary during a project. There are two situations in which this step in CM is applicable. First, there are optional ways to solve a problem, and second, you have to make sure that any adjustments are in line with the project overall goals. With the program manager's and customer's consent, it is the responsibility of SE to put up such a procedure. In such a process, the following components are crucial for evaluation:

- The cost impact, in both budget and schedule.
- The technical significance.
- The broad implications of the change with respect to the system aspect of the program.
- What parts of the organization will be effected by the change?
- Who should have an opportunity to review and evaluate changes?
- How will changes be documented?
- How will the necessary information be distributed to the interested departments of the organization [18]?

Configuration Status Accounting CM is the practice of managing a product and its parts throughout the course of the product's lifecycle. Controlling changes to the product both during development and maintenance; configuration status accounting (CSA), traditionally seen as consisting of recording and reporting the information obtained from utilizing other CM functions; and unique identification of the components within the product structure are all responsibilities of CM. The management of component versions and builds, as well as derivatives of component versions produced by various types of tool processing, has been the main focus of recent CM research and tool development [16]. For example, industrial-scale manufacture of specialized embedded systems requires new CM approaches. It differs from the development time CM that is often supported in that there may be orders of magnitude more variations of the entire product. This is mostly caused by the possibility of individuality in every delivery. The importance of CSA is increased because the new notions raise fresh concerns that need to be addressed. Such tools that support CM as a whole and other software lifecycle tasks have just recently been widely accessible.

The ability to incorporate one's own techniques and tools, as well as support for all phases of software development, is the current trend among these tools. Software engineering environments, or SEEs, are technologies that offer a comprehensive collection of runtime facilities and tools to support the entire software lifecycle. SEEs also offer access to a repository database, a standard user interface, and ways to include new interoperable tools. This research tries to establish a case for improving the process model with CSA to raise the integration level of an SEE. We also aim to encourage the use of CSA for automated product control as part of this. Instead of attempting to offer a comprehensive SEE solution with a database and complex tool integration methods, we want to offer a process integration strategy that enables simple and efficient usage of current tools within controlled workflows [19].

Configuration Verification The process of developing software is becoming more and more dependent on formal software verification. It is on the verge of becoming a natural companion to well-established testing techniques as it develops into a tool with considerably more general use. For instance, the Static Driver Verifier program from Microsoft analyzes the C code of Windows device drivers for compliance with the Windows Driver Model automatically. Complete software systems, on the other hand, include multiple levels of versioning in addition to many source code files. These additional files, which often do not factor into formal verification for low-level programs, include makefiles, configuration files, and shell scripts. Another source of complication for software testing and verification is program configuration. Analyzing (up to) 2n specific configuration instances is necessary for a configuration space with n features that can be independently turned on and off. Additionally, because the collection of potential program executions is now too small for conventional software testing to cover all of them, increasing it exponentially could render the testing methodology useless. Model Checking and other software verification techniques have the ability to cover every program execution scenario, but they are now insufficient—or even sound—because settings are not taken into account.

Lifting is a new method that incorporates configuration data into a traditional verification process. This paper introduces this innovative methodology, focusing on how to use it with build setups and conditional compilation methods. Lifting is the automatic transformation of a software system that can be configured so that all configuration operations are carried out during program execution. In a generative software build process, lifting enables standard verification techniques to reason about the following typical problem areas: D1. There might be a problem with the feature model: D2. It is possible that the limitations of the feature model do not correspond to those imposed by the actual, implemented features: D3. Any variant may fail to comply with software specifications that ought to apply to all variants, such as by resulting in runtime problems. We have selected CBMC, a SAT-based source code bounded model checker for C, as the verification backend [20].

Development

Design Iteration Road Map, Delineation of Different Options Explored New products have a long history of failing in the market because of shifting market dynamics and a lack of awareness of user and customer needs. The ability to trace each of these iterations from customer to origin is important for learning and future development. At least as early as 1973, when Project Sappho was launched, research began to document factors influencing product success and failure.

- Identified the main cause of new product failure as a lack of understanding of user demands. This result has been confirmed throughout time.

- Indicating that little was improving in how well companies were able to incorporate user and consumer insight into the creation of products and their distribution to the market.

- This deficiency prompted exhortations in popular literature.

- A mechanism for businesses to focus more on their customers.

In the meantime, the focus of competition for businesses has evolved from mining raw materials to producing things to providing services, and it is now more and more concentrated on staging client experiences with the ultimate goal of driving consumer transformations. Five companies searched for tools, techniques, and strategies to become more "customer focused" or "market driven" as this transformation took place.

However, being more "customer focused" extends beyond conducting better market research, utilizing QFD, or employing agile development approaches, which aspire to more closely link consumer requirements to options for features and functions. Strategic planning must incorporate all aspects of a customer experience design, not simply those related to the particular specifications of the good or service being provided. The speeding up of technical change is another factor contributing to the requirement for customer attention. For more than 30 years, the fundamental building elements of technology—computing power, bandwidth, and storage—have been advancing exponentially.

We are now just past the knee of the exponential curve. Combinations of these technological building blocks are opening up previously unimaginable prospects as they develop. Examples include driverless vehicles, AI, blockchain, 3-D printing, and other technologies. A business can no longer become dependent on a certain technology and then profit from it. Instead, businesses now need to concentrate on the customer experience (or change) they want to achieve, and then quickly embrace or use the technologies that let them do so. Some people refer to the difficulty of utilizing the right technologies to provide meaningful client experiences as a "wicked dilemma."

There are many parties involved, there is no one right solution, and the only way forward is through iteration, learning from mistakes, and trying again. For instance, the introduction of autonomous vehicles necessitates attention to not only technology development but also the effects on infrastructure design, insurance management, and more. Companies must adopt new ways of thinking, new organizational structures, and new methods of decision-making and strategy-making to set their strategic orientation in such a situation. This includes developing innovative solutions to consumer problems.

In this post, we especially discuss road mapping for portfolio planning and management and how it must adapt to our Volatile, Uncertain, Complex, and Ambiguous (VUCA) reality as one piece of tackling "wicked challenges."

Understanding the Current State of Road Mapping We gained a thorough understanding of current road mapping practice, insight into the ways that businesses are using road maps today, how they are coping with VUCA challenges, and how a "design road mapping" framework can change how a company integrates customer experience design with product or technology choices through in-depth interviews, observations, and experiments. In the first wave of our study, we looked at how road mapping is now done with a focus on where and how understanding of consumers and users is engaged. To accomplish this, we spoke with product managers, technology managers, designers, and design researchers at businesses in the networking, communications, and IT, security solutions, software, e-commerce, financial, online education, Internet, home automation, and healthcare sectors. (Quotes throughout the manuscript are coded with the organizations' product manager (P), technology manager (T), and designer (D) responsibilities for interview respondents). The questions that were covered in these semi-structured interviews included:

- What kinds of road maps does your organization maintain?
- What aspects of the existing road mapping method are effective?
- What is ineffective?
- Who are the key stakeholders in the creation of the road map?
- What improvements have they made to the road maps?
- What more should the road maps include, in your opinion? [21]

Test Equipment, Product, and Artifacts We spent a considerable amount of time investigating development environments. A number of languages and environments were researched because object-oriented design appeared promising. It was crucial to create a workspace that enabled rapid prototyping and testing of user interfaces to meet the "same look and feel goals." Smalltalk-80 was used to start the user interface prototyping process. It was discovered that Smalltalk is a particularly potent environment for rapid prototyping. The user interface simulator and prototype were operational pretty rapidly. Although promising, the Smalltalk development environment was deemed unsuitable for embedded devices. Smalltalk was too complex, too sluggish, and it would be challenging to overcome the linguistic challenges involved in putting Smalltalk into ROM. Because of its at-the-time extremely constrained class libraries and development environment, C++ was considered, but ultimately rejected. Goal C1 seemed to have potential. Many of the benefits of the Smalltalk environment were present, including object paradigms, straightforward object syntax, polymorphism, easy class libraries, and adequate debugging tools. It was anticipated that Objective C would be more compact, quick, and straightforward to integrate into ROM than Smalltalk [22].

Verification and Validation

Modeling

Systems thinking is the foundation of the SE perspective. Systems thinking involves discovery, learning, diagnosis, and dialogue that result in sensing, modeling, and discussion of the real world to comprehend better, define, and engage with systems.

Sensitivity analysis, which examines the connections between outcomes and their probabilities to determine (a) how "sensitive" a decision point is to varying numerical values, is another technique used in decision analysis. (b) Value of information techniques, wherever investing some time in data analysis and modeling might raise the best possible outcome. (c) Multi-attribute Utility Analysis, a technique for establishing equivalencies among disparate units of measurement.

Language for Modeling Systems (SysML) To provide standard representations with clearly specified semantics that can facilitate model and data transfer, the SysML standard is an addition to the family of UML-based standards. A cooperative project between INCOSE and the Object Management Group led to the creation of SysML.

Simulation NPD nowadays makes heavy use of IT instruments such as virtual simulation tools. The main motivation for introducing virtual simulation tools in NPD is to speed up development and lower its cost. Virtual simulation tools, however, do much more. They introduce profound changes in the organization, including the nature of problem-solving, bearing the potential to increase NPD performance beyond cost and lead time reduction.

Virtual simulation tools now play a very important role in NPD. They have been widely hailed to significantly cut development time and costs. Accordingly, virtual simulation tools are often introduced in NPD to reap precisely those benefits. In the academic literature, virtual experimentation is also often regarded as a way to overcome the cost and time limitations of physical experimentation methods. Limiting virtual simulation tools to such considerations, however, misses an important point. In this article, we argue that the contribution of virtual tools to experimentation goes well beyond the incremental improvement of the results obtained with physical experimentation [23].

Prototypes (Increasing Sophistication and Capability)

The design and development phase specifically comprises design concepts via design sketches, full engineering designs, thorough virtual and physical prototypes, design analysis, and prototype testing. Many choices must be discussed about topics like the design criteria, specifications, detailed design and engineering, and the prototype plan during the design and development process. The development and testing of virtual and physical prototypes, as well as the design and sourcing of tools and manufacturing processes, take place during the commercialization phase. These choices frequently involve a lot of interdependencies and call for input from many functional areas across the company [24].

VR and Manufacturing

VR is a technology that is gaining popularity quickly and whose usability is constantly improving. While some VR applications are already developed, others are only getting started. By accelerating the process and enhancing quality and usability, VR opens up new opportunities in the realm of product development. Three of the most common uses of VR that offer beneficial chances for product development, in general, are communication with distant coworkers or other parties, skill training, and simulation (e.g., users, experts, and customers). Before final verification using real prototypes is carried out, prototypes and products can be tested "virtually" (referred to as virtual prototyping) using VR technology.

Before the items are really created, customers can "virtually" test and train how to use them, which can improve usability and ergonomic design. The layout plans (e.g., for a factory, hospital, or city) and features of a product idea can be interactively modified by users and customers using VR. The author advises that brain-aided design and pencil-aided design (PAD), possibly followed by model-aided design (MAD), should most commonly be carried out before VR is adopted, especially for the development of intricate and even complex radical new goods. Computer-aided design (CAD) should ideally be employed initially for the development of unique products after modeling with MAD and/or VR has contributed. When CAD files already exist for re-engineering, PAD is still helpful, but MAD is frequently not required because VR, in that instance, uses existing CAD files, eventually when reverse engineering or scanning of paper designs has been completed [25].

AR and Manufacturing

Systems that overlay computer-generated information over the real world are referred to as AR. This multimodal combination could involve adding virtual annotations to an image, picking up on and amplifying subtle sounds, or using haptic feedback to improve touch sensitivity. AR does not replace the environment like VR does; rather, it improves it. According to Milgram's theory number 1 of a continuum between reality and virtuality, VR technology is frequently utilized in the early stages of the lifecycle of a production system, but AR is primarily used in the control and maintenance phase. It is impossible to maintain a precise division. By seamlessly fusing computer-generated texts, images, animations, and other media with the real world, AR technology gives users a natural interaction experience. The user is presented with an enhanced reality with features created for both the intended application and the end-user interaction experience, resulting in effective perception and creation for them. The use of AR technology has expanded from being confined to science fiction films to a variety of different businesses.

The need for effective methodologies and tools for planning complex production systems has increased because of the shortening of development cycles. The functions of factory planning have recently been added to immersive VR technologies. Because of this, planning timeframes have been cut down, and the quality of the planning outcomes has improved. The development of many virtual planning systems aims to fully integrate all planning processes and necessitates a natural interface with intricate computer models of machinery, factory layouts, etc. The production system must be completely modeled because existing methods and tools can only represent planning objects virtually. The potential advantages of these instruments and this technology, in general, are diminished by their high costs. The AR technology opens up new opportunities for developing the industrial planning process. Virtual planning objects can be layered in a real-world production environment using AR technology. Thus, tasks related to planning can be assessed without modeling the environment surrounding the production site [26].

Levels

Component The crucial phase of the component design process is component-level design. This crucial stage of component design outlines how each component interface, algorithms, data structure, and communication strategies are developed to completely achieve the functionality which is intended for it. Compared to architectural design, component-level design assigns each component function on a more detailed level. Consider a straightforward birthday cake with frosting all over it as the general functionality of a component that is created during the architectural design stage, and a message piped on top as the particular functionality that is implemented during component-level design.

Subsystems Subsystem/component level, which initially generates a set of subsystem and component product performance descriptions, followed by a set of corresponding comprehensive product characteristic descriptions necessary for production.

System System level, which generates a description of the system in terms of performance requirements.

Controlled Environment Operations Operations that are safe to undertake in an environment and that can be managed. This includes vehicle testing on closed circuit tracks for example.

Live (Fire) Exercise Live fire exercises are vehicle exercises in contrived scenarios where there is much less control over the stimuli to which the vehicle may be subjected (Figure 8.7). At the point the product and systems undergo this testing, the product and system capability must be well understood.

FIGURE 8.7   Live exercises are used to learn about the systems as used in the real world.

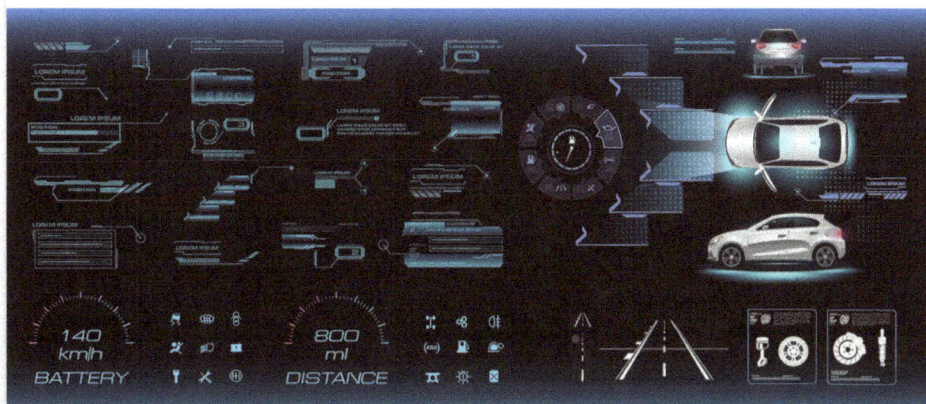

PALERMO89/Shutterstock.com.

References

1. Walden, D.D., Roedler, G.J., and Forsberg, K., "INCOSE Systems Engineering Handbook Version 4: Updating the Reference for Practitioners," Paper presented at in *the INCOSE International Symposium*, Detroit, MI, 2015.

2. Srivastava, A., Abbas, S.Q., and Singh, S., "Enhancement in Function Point Analysis," *International Journal of Software Engineering & Applications (IJSEA)* 3, no. 6 (2012): 129.

3. Bar-Yam, Y., "When Systems Engineering Fails-Toward Complex Systems Engineering," Paper presented at the in *SMC'03 Conference Proceedings. 2003 IEEE International Conference on Systems, Man and Cybernetics. Conference Theme-System Security and Assurance (Cat. No. 03CH37483)*, Washington, DC, 2003.

4. Pedro, O.D.S.L.J., Farias, P.M., and Belchior, A.D., "A Fuzzy Model for Function Point Analysis to Development and Enhancement Project Assessments," *CLEI Electronic Journal* 5, no. 2 (1999): 1-14.

5. Halme, J. and Aikala, A., "Fault Tree Analysis for Maintenance Needs," *Journal of Physics: Conference Series* 364 (2012): 012102.

6. Suma, V. and Nair, T.G., "Effectiveness of Defect Prevention in IT for Product Development," Paper presented at in *the National Conference on Recent Research Trends in Information Technology*, Bangalore, India, 2008.

7. Sharma, P. and Singh, A., "Overview of Fault Tree Analysis," *International Journal of Engineering and Research Technology (IJERT)* 4, no. 3 (2015): 337-340.

8. Tai, A.T., Walter, C.J., Fesq, L.M., and Day, J.C., "Fault-Class-Aware Fault Tree Generation and Analysis," Paper presented at in *the 2013 IEEE International Symposium on Software Reliability Engineering Workshops (ISSREW)*, Pasadena, CA, 2013.

9. Freitas, M.d., Fantinato, M., Sun, V., Thom, L.H. et al., "Function Point Tree-Based Function Point Analysis: Improving Reproducibility Whilst Maintaining Accuracy in Function Point Counting," Paper presented at in *the International Conference on Enterprise Information Systems*, Crete, Greece, 2019.

10. de Freitas Junior, M., Fantinato, M., and Sun, V., "Improvements to the Function Point Analysis Method: A Systematic Literature Review," *IEEE Transactions on Engineering Management* 62, no. 4 (2015): 495-506.

11. Ahmed, F., Bouktif, S., Serhani, A., and Khalil, I., "Integrating Function Point Project Information for Improving the Accuracy of Effort Estimation," Paper presented at in *the 2008 Second International Conference on Advanced Engineering Computing and Applications in Sciences*, Valencia, Spain, 2008.

12. Fraternali, P., Tisi, M., and Bongio, A., "Automating Function Point Analysis with Model Driven Development," Paper presented at in *the Proceedings of the 2006 Conference of the Center for Advanced Studies on Collaborative Research*, Toronto, ON, Canada, 2006.

13. International Function Point Users Group (IFPUG), "Function Point Counting Practices Manual Release 4.3.1," January 2010.

14. Griffin, A., "PDMA Research on New Product Development Practices: Updating Trends and Benchmarking Best Practices," *Journal of Product Innovation Management* 14, no. 6 (1997): 548-551, doi:https://doi.org/https://onlinelibrary.wiley.com/doi/10.1111/1540-5885.1460429.

15. Sheard, S.A. and Mostashari, A., "Principles of Complex Systems for Systems Engineering," *Journal of the International Council on Systems Engineering* 12, no. 4 (2009): 295-311.

16. Cheng, M. and Simmons, J., "Traceability in Manufacturing Systems," *International Journal of Operations & Production Management* 14, no. 10 (1994): 4-16, doi:https://doi.org/10.1108/01443579410067199.

17. Chiocchio, S., Martin, E., Barabaschi, P., Bartels, H.W. et al., "System Engineering and Configuration Management in ITER," *Fusion Engineering and Design* 82, no. 5-14 (2007): 548-554.

18. Pruitt, W.B., "The Value of the System Engineering Function in Configuration Control of a Major Technology Project," *Project Management Journal* 30, no. 3 (1999): 30-36.

19. Viskari, J., "A Rationale for Automated Configuration Status Accounting," in Estublier, J. (Ed.), *Software Configuration Management* (Berlin, Germany: Springer, 1993), 138-144.

20. Post, H. and Sinz, C., "Configuration Lifting: Verification Meets Software Configuration," Paper presented at in *the 2008 23rd IEEE/ACM International Conference on Automated Software Engineering*, L'Aquila, Italy, 2008.

21. Kim, E., Beckman, S.L., and Agogino, A., "Design Roadmapping in an Uncertain World: Implementing a Customer-Experience-Focused Strategy," *California Management Review* 61, no. 1 (2018): 43-70.

22. Dotts, A. and Birkley, D., "Development of Reusable Test Equipment Software Using Smalltalk and C," Paper presented at in *the Addendum to the Proceedings on Object-Oriented Programming Systems, Languages, and Applications (Addendum)*, Vancouver, BC, Canada, 1992.

23. Becker, M.C., Salvatore, P., and Zirpoli, F., "The Impact of Virtual Simulation Tools on Problem-Solving and New Product Development Organization," *Research Policy* 34, no. 9 (2005): 1305-1321.

24. Marion, T.J. and Fixson, S.K., "The Transformation of the Innovation Process: How Digital Tools Are Changing Work, Collaboration, and Organizations in New Product Development," *Journal of Product Innovation Management* 38, no. 1 (2021): 192-215.

25. Ottosson, S., "Virtual Reality in the Product Development Process," *Journal of Engineering Design* 13, no. 2 (2002): 159-172, doi:10.1080/09544820210129823.

26. Doil, F., Schreiber, W., Alt, T., and Patron, C., "Augmented Reality for Manufacturing Planning," Paper presented at the in *Proceedings of the Workshop on Virtual Environments 2003*, San Diego, CA, 2003.

Index

About the Authors

Jon M. Quigley PMP (204278) CTFL is a principal and founding member of Value Transformation LLC, a product development (from idea to product retirement) and cost improvement organization established in 2009. Jon has an Engineering Degree from the University of North Carolina at Charlotte, two master's degrees from the City University of Seattle, and two globally recognized certifications. Jon has more than thirty years of product development and manufacturing experience, ranging from embedded hardware and software to verification and process and project management.

Jon has won awards such as the Volvo-3P Technical Award in 2005 going on to win the 2006 Volvo Technology Award. Jon has secured seven US patents and several international patents. These patents range from multiplexing systems and human-machine interfaces to telemetry systems and driver's aides. Jon has been on the Western Carolina University Master of Project Management Advisory Board as well as the Forsyth Technical Community College Advisory Board. He has also been a guest lecturer at Wake Forest University's Charlotte NC campus as well as at Eindhoven University of Technology (Holland).

Jon has authored or contributed to more than 15 books on product development and project management topics. The books he writes are used in bachelor and master-level classes at universities across the globe, including the Eindhoven University of Technology, Manchester Metropolitan University, San Beda College Manila in the Philippines, and Tecnológico de Monterrey.

In addition to more than 70 different magazines, e-zines, and other outlets. He has recurring podcasts at Tom Cagley's Software Process and Measurement Cast (SPaMCAST) and writes four recurring columns:

1. PMTips Quigley and Lauck's Expert Column
2. *Assembly Magazine*, P's and Q's on project management and quality
3. Automotive Industries, Quigley's Corner on automotive product development
4. MPUG – Microsoft Project User Group Column – Transformation Corner

Jon has given numerous presentations at technical conferences on a variety of domains of product development, including product testing, learning, agile, and project management. He has also frequently been interviewed by many business and project magazines, podcasts, and webinars.

Jon is the co-author or contributed to the following books:

1. Co-author of Taylor & Francis book *Project Management of Complex and Embedded Systems: Ensuring Product Integrity and Program Quality.* ISBN 1420070256. October 21, 2008
2. Co-author of Taylor & Francis book *Scrum Project Management.* ISBN 1439825157. August 16, 2010

3. Co-author of Taylor & Francis book *Testing of Complex and Embedded Systems.* ISBN 1439821402. December 7, 2010

4. Co-author of *Software Test Professionals / Redwood Collaborative Media Professional Development Series – Saving Software with Six Sigma.* ISBN 978-0-9831220-0-5.

5. Co-author of *Software Test Professionals / Redwood Collaborative Media Professional Development Series – Aggressive Testing for Real Pros.* ISBN 978-0-9831220-1-2.

6. Co-author of Taylor & Francis book *Total Quality Management for Project Managers.* ISBN 978-1-4398-85055-5. August 28, 2012

7. Co-author of Taylor & Francis book *Reducing Process Costs with Lean, Six Sigma, and Value Engineering Techniques.* ISBN 978-1-4398-8725-7. December 12, 2012

8. Co-author of Society of Automotive Engineering book *Project Management for Automotive Engineers: A Field Guide.* eISBN PDF 978-0-7680-8315-6; eISBN prc 978-0-7680-8316-3 eISBN epub 978-0-7680-8317-0. August 31, 2016

9. Co-author of Taylor & Francis book *Configuration Management: Theory, Practice and Application.* ISBN 978-1482229356. April 16, 2015

10. Contributor to the e-book *Opening the Door: 10 Predictions on the Future of Project Management in the Professional Services Industry* with MavenLink and ProjectManagers.net

11. Co-author, Taylor & Francis book *Configuration Management, Second Edition: Theory and Application for Engineers and Managers.* July 22, 2019

12. Co-author, Taylor & Francis book *Continuous and Embedded Learning for Organizations.* July 2, 2020

13. Co-author, Taylor & Francis book under contract *Principles, Processes and Practices of Risk Management.* December 2022

14. Co-contributor of Scrum Project Management topic for the *Encyclopedia of Software Engineering.* ISBN 1-4200-5977-7 and e-ISBN 1-4200-5978-5

15. Contributor to the e-book *The Project Manager Who Smiled* by Peter Taylor. Catalog record for the book is available from the British Library ISBN 978-0-9576689-0-4 production June 2013

16. Co-author of Society of Automotive Engineering book "Modernizing Product Development Processes: Guide for Engineers" production August 31, 2023

17. Author of the Society of Automotive Engineering book "Dictionary of Test and Verification" production under contract

18. Co-author of the Society of Automotive Engineering book "Dictionary of Electric Vehicles" production under contract

19. Co-author of the Society of Automotive Engineering book "Dictionary of Commercial Vehicles" production under contract

20. Co-author of the Society of Automotive Engineering book "Dictionary of Automated and Connected Vehicles" production under contract

Jon enjoys the beauty of nature, hiking in the woods, and playing the bass. He lives in Lexington North Carolina with his wife Nancy and son Jackson. Email Jon at jon.quigley@valuetransform.com.

Amol Gulve M.E., PMP is a Chartered Engineer and a Fellow, recognized in the field of mechanical engineering as an eminent engineer. Amol has a master's degree from the University of Detroit Mercy and a bachelor's degree from the University of Mumbai. Amol has been working in the automotive and commercial heavy-duty truck industry for more than seventeen years and has a strong background in product development, process optimization, quality, continuous improvement, embedded engineering, and manufacturing.

Amol was recently awarded the SAE Forest R. McFarland Award for his outstanding contributions toward the work of the SAE activities focused on the dissemination of technical information, editorial management, and standards development. Amol is also honored as a Fellow and a Chartered Engineer by the Institution of Mechanical Engineers, which is the highest level of recognition for individuals who have demonstrated commitment and contribution to engineering. Amol is also an active member of the Institute of Electrical and Electronics Engineers as a senior member and Sigma Xi honor society where he is involved in technical discussions and judging activities to develop young engineers. Tau Beta Pi, the oldest engineering honor society, has recognized Amol as Eminent Engineer for his history of academic achievements and commitment to personal and professional integrity.

Amol has authored or contributed to several technical papers and is also actively engaged as an editorial manager and guest editor for international journals. The papers he has published have focused on key technical areas of interest in the automotive industry such as vehicle rollovers, safety, reducing CO_2 emissions, and modularity strategy. He is regularly invited by universities across the globe to participate as a design judge and guest speaker and provide mentoring to students for career opportunities.

Amol has authored or contributed to the following papers:

1. Characterization of Key Vehicle Parameters Affecting Vehicle Rollover Using Adams™ Simulations Software and 1/10th Scale Model Testing (SAE # 2006-01-1951)

2. The Transport Hierarchy: A Cross-modal Strategy to Deliver a Sustainable Transport System (IMechE Policy Paper)

3. Modular Approach to Developing Platform Solutions across Multiple Brands and Segments (SAE # 2021-01-0837).

Amol enjoys biking, playing tennis, reading, and volunteering. Amol recently moved to Dallas, Texas with his wife Amruta and two daughters Aisha and Navya to work for a renowned heavy-duty truck manufacturer to lead the future of transportation. Email Amol at agulve@gmail.com.